Student Manual to Accompany

AN INTRODUCTION TO FIRE PROTECTION

Deborah Olson

Delmar Publishers
an International Thomson Publishing Company I(T)P®

Albany • Bonn • Boston • Cincinnati • Detroit • London • Madrid
Melbourne • Mexico City • New York • Pacific Grove • Paris • San Francisco
Singapore • Tokyo • Toronto • Washington

NOTICE TO THE READER

Publisher does not warrant or guarantee any of the products described herein or perform any independent analysis in connection with any of the product information contained herein. Publisher does not assume, and expressly disclaims, any obligation to obtain and include information other than that provided to it by the manufacturer.

The reader is expressly warned to consider and adopt all safety precautions that might be indicated by the activities described herein and to avoid all potential hazards. By following the instructions contained herein, the reader willingly assumes all risks in connection with such instructions.

The publisher makes no representations or warranties of any kind, including but not limited to, the warranties of fitness for particular purpose or merchantability, nor are any such representations implied with respect to the material set forth herein, and the publisher takes no responsibility with respect to such material. The publisher shall not be liable for any special, consequential or exemplary damages resulting, in whole or in part, from the readers' use of, or reliance upon, this material.

Cover photo courtesy of George Hall/Code Red

Delmar Staff

Publisher: Alar Elken
Acquisitions Editor: Mark Huth
Developmental Editor: Jeanne Mesick
Production Coordinator: Toni Bolognino
Art and Design Coordinator: Cheri Plasse

COPYRIGHT © 1998
By Delmar Publishers
an International Thomson Publishing Company

The ITP logo is a trademark under license

Printed in the United States of America
For more information, contact:

Delmar Publishers
3 Columbia Circle, Box 15015
Albany, New York 12212-5015

International Thomson Publishing Europe
Berkshire House 168-173
High Holborn
London, WC1V7AA
England

Thomas Nelson Australia
102 Dodds Street
South Melbourne, 3205
Victoria, Australia

Nelson Canada
1120 Birchmount Road
Scarborough, Ontario
Canada M1K 5G4

International Thomson Editores
Campos Eliseos 385, Piso 7
Col Polanco
11560 Mexico D F Mexico

International Thomson Publishing GmbH
Königswinterer Strasse 418
53227 Bonn
Germany

International Thomson Publishing Asia
221 Henderson Road #05-10
Henderson Building
Singapore 0315

International Thomson Publishing - Japan
Hirakawacho Kyowa Building, 3F
2-2-1 Hirakawacho
Chiyoda-ku, 102 Tokyo
Japan

All rights reserved. No part of this work covered by the copyright hereon may be reproduced or used in any form or by any means—graphic, electronic, or mechanical, including photocopying, recording, taping, or information storage and retrieval systems—without written permission of the publisher.

1 2 3 4 5 6 7 8 9 10 XXX 02 01 00 99 98 97

Library of Congress Cataloging-in-Publication Data 96-11630

ISBN 0-8273-8229-4

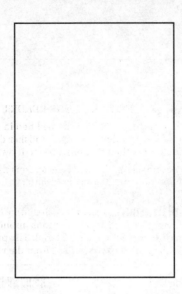

CONTENTS

	Preface	vii
Chapter 1	**Fire Technology Education and the Firefighter Selection Process**	**1**
	Introduction	2
	Questions	2
	Exercises	2
	Assignments	12
Chapter 2	**Fire Protection Career Opportunities**	**15**
	Introduction	16
	Questions	16
	Exercises	17
	Assignments	19
Chapter 3	**Public Fire Protection**	**21**
	Introduction	22
	Questions	22
	Exercises	24
	Assignments	26
Chapter 4	**Chemistry and Physics of Fire**	**29**
	Introduction	30
	Questions	30
	Exercises	31
	Assignments	35
Chapter 5	**Public and Private Support Organizations**	**39**
	Introduction	40
	Questions	41
	Exercises	42
	Assignments	44

Chapter 6	**Fire Department Resources**	**47**
	Introduction	48
	Questions	48
	Exercises	50
	Assignments	54

Chapter 7	**Fire Department Administration**	**57**
	Introduction	58
	Questions	58
	Exercises	59
	Assignments	61

Chapter 8	**Support Functions**	**63**
	Introduction	64
	Questions	64
	Exercises	65
	Assignments	70

Chapter 9	**Training**	**71**
	Introduction	72
	Questions	73
	Exercises	74
	Assignments	78

Chapter 10	**Fire Prevention**	**81**
	Introduction	82
	Questions	82
	Exercises	83
	Assignments	88
	Excerpts from NFPA 901—Uniform Coding	89

Chapter 11	**Codes and Ordinances**	**103**
	Introduction	104
	Questions	104
	Exercises	106
	Assignments	107

Chapter 12	**Fire Protection Systems and Equipment**	**109**
	Introduction	110
	Questions	110
	Exercises	113
	Assignments	114

Chapter 13	**Emergency Incident Management**	**117**
	Introduction	118
	Questions	118
	Exercises	120
	Assignments	122

Chapter 14   Emergency Operations   123

     Introduction   124
     Questions   124
     Exercises   126
     Assignments   130

PREFACE

The information in this manual is designed to reinforce and compliment the textbook *An Introduction To Fire Protection*, by Robert Klinoff. The manual is comprised of chapters that correspond to the same numbered chapter in the textbook. There are three parts to each section. They are Questions, Exercises, and Assignments.

The questions relate to and reinforce the information in the text book. The exercises give the student the opportunity to complete forms that are discussed in the textbook and relate the information they are learning to case studies of actual incidents or common scenarios. The assignments direct the student out into the real world. Flexibility is given to the student to complete the assignments as they relate to the local fire department or the fire department in the jurisdiction of their choice. This allows the students to become more knowledgeable about the fire department in which they are most interested and how they compare to departments that are selected by other students. By completing the assignments, individually or as a group, the student will develop a better understanding of the fire service as he or she relates information in the text to the actual practice of fire protection. The assignments may also reveal areas within a fire department where the student would be able to participate in training or volunteer activities if desired.

Fire Technology Education and the Firefighter Selection Process

"It is thrifty to prepare today for the wants of tomorrow."
Aesop—The Ant and the Grasshopper

INTRODUCTION

With the high level of competition for jobs in the fire service, being well prepared for each step of the testing process is a necessity. The questions, exercises, and assignments that follow are designed to give you a better understanding of the testing process and help you prepare for it.

Questions

1. It has been proven that even though people have been trained to perform in a certain way, it is difficult to predict what they will do when in danger or in an emergency situation.

    True   False

2. Each year, approximately 100 firefighters die in the line of duty.

    True   False

3. Appearance is not important during the oral interview since the panel is evaluating the applicant's education and work experience.

    True   False

4. The Job Announcement is a valuable resource containing important information that can be utilized when filling out the Job Application.

    True   False

Match each term on the right to the statement that best describes it.

_____ 5. A set number of core courses and electives to achieve a specified number of college credits.

_____ 6. Training that includes operating equipment, laying hose lines, and raising ladders.

_____ 7. A set of core courses and electives combined with classes to meet graduation.

_____ 8. Training that includes hydraulics, pump theory, and equipment capabilities.

A. Manipulative training

B. Technical training

C. Associate Degree in Fire Technology

D. Fire Technology Certificate

Exercises

The exercises in this section consist of filling out a job application using information from the hiring announcement, completing a basic math review, and answering oral interview questions.

Exercise 1

The Hiring Announcement and Job Application. Read all of the instructions on both the hiring announcement and the job application carefully. Some applications must be completed in ink or typed. Others, like the USDA Forest Service Application for Temporary Employment, are read by computer and must be filled out using a No. 2 pencil. Note the date or dates on the job announcement specifying when applications will be accepted. Many agencies have a recruitment period and will only accept applications on the specific dates listed. The U.S.F.S. has a National Recruitment period each year for temporary employment; whereas, city and county department recruitments are often on an as needed basis.

Complete the job application using the hiring announcement. On the application, specify how past employment and experience relate to the duties in the hiring announcement. Include copies of certificates, and if needed, a separate sheet of paper listing additional training or education. Completing this exercise will provide you with a valuable resource and reference when applying for a position in the fire service.

City of River View
Employment Opportunity

Apply To: Personnel Department
1000 Main Street
River View
Phone: 555-0555

FIREFIGHTER

SALARY: $2,000 to 2,500 monthly

NOTE: Employment applications must be completed in accordance with instructions on the face of the application form. Incomplete applications will be subject to rejection.

QUALIFICATIONS: Graduation from high school or successful completion of a general development test indicating high school level. Proof of G.E.D. must be submitted with application. Possession of a valid motor vehicle operator's license.

NOTE: As a condition of employment, the appointee shall not use tobacco products during his or her tenure with the River View City Fire Department, either on or off duty.

PRE-EMPLOYMENT PHYSICAL EXAM:

MEDICAL: Includes spine and chest x-rays, blood chemistry and electrocardiograph tests. Weight must be proportionate to height and age.

VISION: 20/50 uncorrected in each eye, no color deficiency, no permanent or progressive eye abnormalities.

HEARING: No hearing deficiency.

AGE LIMITS: Applicants must have reached their 18th birthday at the time of final filing date for receiving applications.

DUTIES: Under supervision, responds to alarms and assists in suppression of fires; cleans up and performs salvage operations after fires; assists in maintaining and caring for fire apparatus, equipment, fire station and grounds; responds to emergency calls, operates resuscitator and administers first aid; makes residential and business inspections to discover and eliminate potential fire hazards and to educate the public in fire prevention; participates in continuing training and instruction program by individual study of technical material and attendance at scheduled drills and classes; may drive and operate fire trucks and similar fire equipment; assists in roadside burning and in building or clearing fire breaks or fire roads; may train or assist in training auxiliary firefighters; may act as relief for driver/operator or company officer; and does related work as required.

WRITTEN TEST: Consists of components testing for the ability to apply information obtained from written materials; to perform basic calculations by using algebraic, arithmetic, and geometric principles; to determine and apply mathematical principles necessary to provide needed information; to interact with the public in a positive manner; to perform routine cleanup and maintenance duties under non-emergency conditions; and to perform tasks without creating safety hazards to self or others. The written exam will be administered at the River View High School Auditorium, 6501 Juniper Lane, River View, on May 4.

PERFORMANCE SKILLS TEST: Will consist of components testing for: the ability to compose and write correspondence in a manner that is readily understandable by the general public; to remember a small amount of information for a period of time; to quickly recall varied information that is needed to perform a task; to recognize actual or potential problems of a physical nature; to observe and obtain needed information from visual sources; and to apply information attained from visual sources.

INTERVIEW: Will be conducted for the purpose of appraising training, experience, interest, and personal fitness for the position. Will consist of components testing for the ability to speak effectively to groups; to make decisions under emergency conditions; to follow personal hygiene practices; to work according to specific standards, instructions, or procedures; and to make decisions from complete information.

Applicants must attain at least a 70 percent score on each phase of the examination process.

PHYSICAL ABILITY TEST: The River View City Fire Department will schedule, notify and administer a physical ability test after the eligible list has been established. The ability test will be administered in groups of 40, dependent upon the number of vacancies, certifications and appointments that occur.

The names of individuals who do not pass the physical ability test or the pre-placement physical examination will be removed from the Eligible List.

FILING DATE: All applications and additional materials must be received in the office of the River View City Personnel Department not later than 5:00 P.M. on April 13. Applicants are encouraged to be thorough in the materials they submit to be evaluated.

City of River View
Application For Employment
Phone 555-0555

Personnel Department   An Equal Opportunity
1000 Main Street   Employer
River View

POSITION APPLIED FOR: _____

Read the job bulletin to determine if you meet the requirements. Type or print clearly in dark ink. False statements may be cause for rejection of the application, removal of name from eligible list, or dismissal from position. All information is subject to verification.

LAST NAME   FIRST   MIDDLE INITIAL
_____

ADDRESS   NUMBER   STREET
_____

Check categories you are willing to work:

Full Time _____   Part Time _____   Temporary _____

Weekends & Holidays _____   Nights _____

EDUCATION AND TRAINING

Highest Grade Completed _____ Graduation Date __/__/__ or GED __/__/__

Name & Location of last grade or high school attended _____
_____

Name and Location	Semester Units	Quarter Units	Major Subjects	Dates Attended	Degrees or Certificate

List name and location of trade or vocational schools, colleges, universities, apprentice or training programs attended:

ADDITIONAL INFORMATION

Provide any additional information applicable to this position. Include professional affiliations, volunteer activities, certificates of professional or vocational competence or licenses, or the ability to use specialized tools or equipment related to the job:

_____
_____

(continued)

_____
_____
_____
_____
_____
_____
_____

Attach additional sheet or use back of sheet for additional information.

EXPERIENCE

List all periods of employment and unemployment for the last ten years, starting with the most recent and working back. Start with present employment. Indicate any discharge or forced resignation. List periods of U.S. military service regardless of when it occurred. Give complete information. A resume does not substitute for this section, but may be submitted in addition to it.

From ___/___ to ___/___ Company or Employer's Name & Address: Supervisor's Name & Title: Reason for Leaving:	Title of Your Position: Number of Hours Worked: Duties of Your Position:
From ___/___ to ___/___ Company or Employer's Name & Address: Supervisor's Name & Title: Reason for Leaving:	Title of Your Position: Number of Hours Worked: Duties of Your Position:
From ___/___ to ___/___ Company or Employer's Name & Address: Supervisor's Name & Title: Reason for Leaving:	Title of Your Position: Number of Hours Worked: Duties of Your Position:
From ___/___ to ___/___ Company or Employer's Name & Address: Supervisor's Name & Title: Reason for Leaving:	Title of Your Position: Number of Hours Worked: Duties of Your Position:

Exercise 2

Basic Math Review. Firefighting entrance exams, along with the profession itself, require the ability to perform and apply basic arithmetic operations such as addition, subtraction, multiplication, and division, as well as percentages, algebra, and geometry. The problems that follow represent principles that may be included in the fire department testing process. If necessary, a math class in Introductory Algebra or one beginning at your present mastery level might be helpful.

Write the number in **Expanded Form**:

Example: $132 = 1 \times 10^2 + 3 \times 10^1 + 2$

   1. $25 =$ **2.** $4{,}579 =$ **3.** $136 =$ **4.** $93{,}027 =$

Find the **Sum**:

Example:	**5.**	**6.**	**7.**
324	235	78.2	7,852
253	421	5.7	6,923
700	+343	+42.8	4,687
+582			+3,510
1,859			

   8. $21.5 + 77 + 59.5 + 27 =$ _____

Find the **Difference**:

Example:	**9.**	**10.**	**11.**
97	92	7.24	10,532
−49	−77	−5.98	− 1,341
48			

   12. $11.6 - 5.82 =$ _____

Find the **Product**:

Example:	**13.**	**14.**	**15.**
24	75	39.4	725
× 9	× 13	× 1.2	× 168
216			

   16. $4.69 \times 1.2 =$ _____

Find the **Quotient** to the nearest tenth:

Example: $84 \div 2 = 42$

   17. $268 \div 4 =$ **18.** $550 \div 25 =$ **19.** $3080 \div 2.8 =$ **20.** $11881 \div 15 =$

Evaluate each of the following using the **Order of Operations**:

Example: $12 - 3 \times 2 = 6$

   21. $36 \div 3 \times 2 =$ **22.** $3 \times 4 + 2 \times 5 - 2 \times 0 =$ **23.** $8 \div 2 \times 3 =$

Write the **Prime Factorization** of each number.

Example: $18 = 2 \times 3^2$

   24. 27   **25.** 40   **26.** 31   **27.** 360

Find the **LCM (Least Common Multiple)** of the numbers.

Example: 9,12
 $9 = 3 \times 3$
 $12 = 2 \times 2 \times 3$
 $LCM = 3^2 \times 2^2 = 36$

28. 35,50

29. 8,32,50

Simplify each fraction to lowest terms.

Example: $\dfrac{2}{4} = \dfrac{1}{2}$

30. $\dfrac{24}{32} =$

31. $\dfrac{35}{50} =$

32. $\dfrac{324}{122} =$

Fill in the missing numbers.

Example: $\dfrac{5}{10} = \dfrac{50}{100}$

33. $\dfrac{3}{5} = \dfrac{}{45}$

34. $\dfrac{12}{24} = \dfrac{}{88}$

35. $\dfrac{37}{125} = \dfrac{}{250}$

Evaluate the following, expressing each answer in simplest form.

Example: $\dfrac{5}{8} + \dfrac{1}{8} = \dfrac{6}{8} = \dfrac{3}{4}$

36. $\dfrac{8}{12} + \dfrac{2}{12} =$

37. $\dfrac{7}{15} + \dfrac{8}{30} =$

38. $\dfrac{7}{12} - \dfrac{4}{20} =$

39. $\dfrac{10}{33} \times \dfrac{2}{5} =$

40. $9/10 \div 2/3$

Convert each fraction to a decimal.

Example: $\dfrac{1}{2} = .50$

41. $\dfrac{7}{8} =$

42. $\dfrac{13}{25} =$

43. $\dfrac{5}{200} =$

Convert each fraction to a percent.

Example: $\dfrac{1}{2} = 50\%$

44. $\dfrac{7}{8} =$

45. $\dfrac{13}{25} =$

46. $\dfrac{5}{200} =$

Find the **Square Root** of each number.

Example: $\sqrt{25} = 5$
 $5 \times 5 = 25$
 $5^2 = 25$
 square root = 5

47. $\sqrt{49} =$

48. $\sqrt{121} =$

49. $\sqrt{97} =$

Solve each of the following problems:

Example: The sum of two numbers is 39. The larger number is 3 less than twice the smaller number. Find both numbers.
Let x represent the smaller number
Let 2x – 3 represent the larger number
x + 2x – 3 = 39
3x – 3 = 39
3x = 42
x = 14
If x = 14 then 2x – 3 = 25
The smaller number is 14. The larger number is 25.

50. A rope 20 feet long is cut into two pieces so that one piece is 6 feet longer than the other. How long is each piece?

51. The sum of two numbers is 66. The larger number is 6 more than 3 times the smaller number. Find both numbers.

52. Use the following formula to calculate the capacity in gallons of a water tank that is 5 feet in diameter and 10 feet tall.

    Formula: $C = d^2 6H$

    Where: d = the diameter of the tank

    H = height

53. Use the following formula to determine the nozzle reaction in pounds of a .5 inch diameter tip with a nozzle pressure of 50 p.s.i.

    Formula: $NR = 1.5 d^2 NP$

    Where: NR = nozzle reaction in p.s.i.

    d = diameter of nozzle tip

    NP = nozzle pressure

54. A 500 gallon water tank is being filled at the rate of 36 gallons per hour. There is an open valve on the tank that is discharging 11 gallons of water per hour. If you begin filling the tank at 8:00 A.M. on Wednesday when will the tank be full?

55. A twenty-person hand crew is capable of cutting a 3 foot wide by 660 foot long fire break in one hour. What is the area in square feet they will have cleared in two hours?

56. A city block is 600 feet long by 300 feet deep. There is a fire hydrant located on opposite corners of the block. What is the shortest distance between the two hydrants?

    Where: The two sides of a right triangle (length & depth of block) are known the third side of the triangle can be found.

    Formula: $A^2 + B^2 = C^2$

Exercise 3

Oral Interview. Being prepared for an oral interview is very important. This is when you have the opportunity to sell yourself as the best qualified individual for the position. If you are uncomfortable speaking in front of people, or have trouble putting into words what you want to say, a speech class may be beneficial. After answering the following oral interview questions, practice giving your answers verbally to another person without looking at your written answers. Do not memorize your answers word for word as this can detract from your presentation.

ORAL INTERVIEW QUESTIONS

1. Why are you interested in becoming a firefighter with this department?

   _____
   _____
   _____
   _____
   _____

2. Tell us about your education, work experience, and any additional training you have received.

   _____
   _____
   _____

3. What experience have you had in dealing with the public or groups of people with different social and economic backgrounds? Give specific examples.

4. If you encountered a situation for which there was an established procedure, but felt you could accomplish the task another way more efficiently, would you follow the established procedure or deviate from it? Please explain your answer.

5. What would you do if in the course of a fire operation your supervisor tells you to perform an operation you feel is completely unsafe?

6. What would you do at a fire scene if after your supervisor has assigned you to a task, his supervisor comes by and reassigns you to a different task?

___

7. Do you have any questions for us, or is there anything you would like to add that we have not asked you about?

___

Assignments

1. Talk to a counselor at the local college, or college of your choice, to find out the specific courses required for a certificate in Fire Technology.

___

2. Contact the local fire department, or fire department in the jurisdiction of your choice, to find out the prerequisites for filing a job application for entry level firefighter with them.

___

3. Contact the local fire department, or fire department in the jurisdiction of your choice, to find out what the steps are in their firefighter testing process and of what specifically each of those steps consist. Find out the what-where-when-how of each step. Some areas that may be included in the testing processes are listed below.

Job application and job announcement: _____

_____

_____

Written examination: _____

_____

_____

Skills test: _____

_____

_____

Oral examination(s)/interview(s): _____

_____

_____

Physical ability/agility: _____

_____

_____

Medical examination: _____

_____

_____

Additional steps for department of your choice: _____

_____

_____

_____

4. As a class or in small groups, set up a mock oral interview. Contact local fire department personnel to see if any of them would be willing to assist in the process. Students may rotate from being the applicant to being the interviewer after they have been interviewed.

Fire Protection Career Opportunities

"When men are employed, they are best contented; for on the days they worked they were good-natured and cheerful, and, with the consciousness of having done a good day's work they spent the evening jollily; but on our idle days they were mutinous and quarrelsome."

Benjamin Franklin

INTRODUCTION

Moving into the 21st century, the fire department of today has evolved tremendously from that of its predecessors. Technological advancements in all types of industry, along with increased budget constraints, are requiring fire departments to become involved in all types of public safety emergencies. In many areas, fire department personnel are routinely dispatched to incidents other than fires. These incidents include, but are by no means limited to medical emergencies, rescues, vehicle accidents, building collapse, swift water rescue, train derailments, airplane incidents, flooding, earthquakes, and hazardous material incidents such as clandestine drug labs or hazardous material spills and releases. The evening news attests to the fact that whenever public safety is threatened, the fire department is usually among the first responders.

Non-fire suppression positions often available within a fire department include reviewing new and existing construction for adequate fire and life safety requirements, performing routine checks of fire prevention systems, testing for adequate water systems and supplies, vegetation management, hazard reduction programs, and presenting fire prevention and disaster preparedness education to the communities they serve.

Private industry also employs fire suppression and non-fire suppression support personnel. Many people have gained valuable experience working in fire protection in the private sector; others have had to retire from active fire suppression positions and have gone to work for private industry.

Disneyland has a private fire department. At the Disneyland in Anaheim, California, department personnel consist of a fire chief, two assistant chiefs, 13 full-time firefighters, and 53 part-time firefighters. Many of the part-time firefighters currently work for municipal departments, others are retirees from municipal departments. Disneyland firefighters are responsible for fire protection system maintenance and testing as well as fire suppression in the park. Application requirements for a position with the Disneyland Fire Department include five years of experience with a municipal fire department.

In public fire departments, as well as in private industry, personnel are necessary in both fire suppression and non-fire suppression positions to support the types of incidents and activities that are routinely performed in each sector.

Questions

Match the fire suppression position with the job description that best describes it.

_____ 1. Responds under supervision to fire alarms and emergencies to protect life and property.

_____ 2. Firefighter duties with additional responsibility of advanced life support.

_____ 3. Operates heavy motorized equipment in fire control work and constructs and maintains fire breaks and roads.

A. Firefighter Trainee

B. Probationary Firefighter

C. Firefighter

D. Firefighter Paramedic

E. Fire Heavy Equipment Operator

_____ 4. Assists in fire suppression and prevention under close supervision in a learning capacity.

_____ 5. Responds to alarms, assists in suppression of fires, cleans up and performs salvage operations, assists in maintenance and care of fire apparatus and equipment.

6. List three non-fire suppression positions often found in the public fire service.

   1. _____
   2. _____
   3. _____

7. List three fire protection careers that can be found in the private sector.

   1. _____
   2. _____
   3. _____

8. Private industry, including insurance companies, often hire civilians with fire technology backgrounds for work in loss prevention and safety.
   True   False

9. Ideas and development of new fire fighting and safety equipment is best left to researchers and scientists.
   True   False

10. Seasonal wildland fire fighting experience has no value when applying for a position with a municipal fire department.
    True   False

11. The fire protection system maintenance specialist services fire extinguishers and fixed fire protection systems.
    True   False

Exercises

Exercise 1

Read the scenario and complete the chart that follows.

On March 1, 1996 at 02:30, a fire alarm was received by an alarm company from a fruit packing plant that had an automatic fire alarm system. Personnel from the alarm company contacted the fire department's emergency communication center to report the alarm. At 02:38, three engine companies and a battalion chief were dispatched to the fire. While fire personnel were enroute, the dispatchers were able to contact the safety officer at the plant who confirmed that there was a fire that appeared to be located in the storage area of the plant. Employees were in the process of evacuating the area. The company officer of the first-in engine company requested that the hazardous material team be dispatched because of the potential release of ammonia used for refrigeration.

The fire was located in a hallway that separated the storage rooms from each other. It was extinguished by the first-in engine company consisting of a company officer, driver/operator, firefighter, and probationary firefighter. Because of the suspicious nature of the fire, the scene was secured and an arson investigator was requested. In the following chart:

1. Identify and briefly describe the fire protection job positions that were utilized on this incident.
2. Indicate in which category the position would most likely be found, fire suppression protection or non-fire suppression protection.
3. Indicate if the position would most likely fall under public fire protection or private fire protection.

Position	Fire Suppression or Non-Fire Suppression	Public Protection or Private Protection

Exercise 2

Refer to the job application you completed in the first section of the manual.

1. Which of the fire suppression jobs are you presently qualified to apply for?

   _____

   _____

2. Which of the non-fire suppression jobs are you presently qualified to apply for?

   _____

   _____

3. What additional training or education is required or would make you a more attractive candidate when applying for the career position of your choice?

　_____

　_____

　_____

4. Which job positions could you pursue to gain experience and make you a more attractive candidate when applying for the career position of your choice?

　_____

　_____

　_____

Assignments

1. Contact your local fire department, or the fire department in the jurisdiction of your choice, and find out what type of fire suppression and non-fire suppression positions are available. What are the entrance requirements for those positions?

　_____

　_____

　_____

　_____

2. Contact local companies and facilities, or those in the jurisdiction of your choice that have private sector fire protection. Fire department personnel may be able to tell you which facilities in their districts utilize private fire protection positions and programs. List some of the fire suppression and non-fire suppression positions available along with the entrance requirements for the positions.

　_____

　_____

　_____

　_____

Chapter 3

Public Fire Protection

*"Progress, therefore, is not an accident, but a necessity . . .
It is a part of nature."*

Herbert Spencer

INTRODUCTION

Since the beginning of human history, men and women from around the world have fought side by side to protect themselves and their surroundings from hostile fires. This has continued through the decades as fire suppression has progressed from bucket brigades to the specialized equipment available today. Loss of lives and property to fire has and will continue to be a motivating force in the evolution of public fire protection.

In the early days of fire suppression, everyone in town, men, women, and children were needed to help with the bucket brigades. Those with more strength passed the full buckets of water from the water source to the fire, while those with less strength passed the empty buckets back to the water source.

As towns and cities grew, volunteer fire departments began to emerge and develop equipment that replaced the bucket brigades. Some of the well-known names of people who once were volunteer firefighters include George Washington, Thomas Jefferson, Benjamin Franklin, and Paul Revere. Women who worked as volunteer firefighters in suppression efforts during this time included: Molly, a slave who in 1818 responded to fires with her owner, a member of a New York engine company; Marina Betts, who joined in the fire suppression efforts in Pittsburgh in 1820; and Lillie Hitchcock, who in the mid-1800s helped with fire suppression efforts in San Francisco.[1]

Public and privately funded departments gradually replaced the competitive volunteer departments. The volunteer departments which existing today are organized similarly to those of public and privately funded departments.

The National Fire Protection Agency was organized in 1896 and continues to develop and update standards for all aspects of the fire service. Standard 1201 4-3.2 states that: the Strategic Planning Process shall attempt to project the future fire protection needs of a community for periods of 10 and 20 years.[2] In addition to projecting future needs of the communities they serve, the fire service will continue to evolve as new ways of protecting and preserving life and property become apparent during fireground operations.

Questions

Match the events of the evolving fire service with the correct locations and dates.

_____ 1. First Volunteer Fire Company

_____ 2. Corps of Vigiles

_____ 3. Advent of fire insurance companies

_____ 4. First publicly funded department in America

_____ 5. First recorded building code prohibited wood or plaster chimneys

_____ 6. Legislation requiring new construction to be of brick or stone with roofs of tile or slate

A. Rome, 6 A.D.
B. New Amsterdam, 1648, later named New York
C. London, 1666
D. Boston, 1679
E. Philadelphia, 1736
F. Cincinnati, 1853

7. The demise of the volunteer fire companies was mostly due to
   A. lack of funding.
   B. competition among the volunteer companies.
   C. unavailability of personnel.
   D. lack of training.

8. The effectiveness of the first hand pumpers was increased by
   A. increasing the diameter of the gooseneck.
   B. positioning the pumper closer to the fire.
   C. adding hose to the top mounted discharge.
   D. increasing the size of the water supply tub.

9. Which of the following is not a true statement regarding the first steamers?
   A. They were welcomed by the volunteer firefighters.
   B. They could pump as long as there was coal to fuel them.
   C. They were too heavy to be pulled by hand.
   D. Fewer people were needed to operate them.

10. Which of the statements is not accurate concerning the planning and formulation of a public fire defense system?
    A. Policies and procedures must be established.
    B. System plans are permanently implemented.
    C. Goals form the basis for the system's objectives.
    D. Cost analyses are needed to determine funding requirements.

11. The greatest cost in a full-time fire department is for
    A. personnel.   C. stations.
    B. equipment.   D. training.

12. List 3 goals of fire protection.
    1. _____
    2. _____
    3. _____

13. Private fire protection was offered by insurance companies after the establishment of public fire protection programs.
    True   False

14. Factors that contributed to the evolution of the fire service included large losses of life and property revealing the need for increased fire fighting capabilities.
    True   False

15. Due to increases in technology, building, and industry the current trend in fire protection is to concentrate on increased fire suppression capabilities as the most efficient means of protecting life and property.
    True   False

16. As the fire service has evolved, the importance of firefighter safety has been increasingly recognized.
    True   False

Exercises

Exercise 1

Develop and draw a home emergency plan for your residence.[3] Include the following items, using colored pencils if available.

1. A drawing of your home, to scale, in black.
2. Mark two means of egress from each room. Use green arrows to identify these exits.
3. Mark location of fire extinguishers using a red "FE."
4. Mark location of smoke detectors using a red "SD."
5. Outside of residence mark location of family meeting place using a red "MP."
6. Mark location of gas shut-off with a yellow "G."
7. Mark location of water shut-off with a blue "W."
8. Mark location of electrical panel with an orange "E."

Exercise 2

Read the incident information concerning The Great Boston Fire of 1872 then answer the questions that follow.

One of the last great fires of the nineteenth century occurred in Boston. On the evening of Saturday, November 9th, 1872, a fire started in a building in the business district. When it was finally extinguished 13 people were dead, including two firefighters who were killed when the building in which they were attempting a rescue collapsed. A total of 776 buildings over an area of 60 acres had burned and a property loss of seventy-five million dollars had been calculated. The fire was finally controlled late on Sunday afternoon.

There were several factors that contributed to the loss of life and property from the fire. When the fire chief arrived at the scene of the fire, 10 minutes after the alarm had been received, he found not only the building of origin on fire, but several additional large buildings. Mobilizing forces in a timely manner was difficult due to the fact that only 89 of the 472 member department were permanent. Equipment response was delayed because most of the horses were ill and unable to be utilized. This resulted in the heavy apparatus, hose wagons, and ladder trucks having to be pulled to the fire by volunteers.

Problems in attacking the fire in an area where some buildings had six stories included hose streams that could reach only to the third floor and ladders with a maximum length of 40 feet. Additional firefighters responded from 37 towns, increasing the fire fighting force to 1600 people. Coordinated efforts for water supply were hampered due to the difference in types and threads of hose couplings and fittings from different fire departments. Inadequate water supply lines to the hydrants that were being used to support fire fighting efforts, as well as buildings of combustible material also contributed to fire control problems.[1, 4]

1. What type of fire apparatus would have been used at this fire?

_____

_____

_____

2. List three problems encountered operationally on the fire and briefly describe how they relate to the evolution of fire service apparatus and practices.

   1. _____
      _____
      _____

   2. _____
      _____
      _____
      _____

   3. _____
      _____
      _____

3. List two types of prevention measures that might have evolved from this fire and briefly describe how they would help decrease loss of life and property in the future.

   1. _____
      _____
      _____

   2. _____
      _____
      _____

4. List two types of fire fighting standards which might have evolved from this fire, and briefly describe how they would help decrease loss of life and property in the future.

   1. _____
      _____
      _____

   2. _____
      _____
      _____

Assignments

1. Contact the local fire department, or fire department in the jurisdiction of your choice, to determine the current level of fire protection they provide.

   A. Square miles in department's jurisdiction: _____

   B. Population of department's jurisdiction: _____

   C. Number of stations: _____

   D. Number of apparatus at stations: _____

   E. Number of personnel in an engine company: _____

   F. Level to which department personnel are trained including fire, medical, rescue, and hazardous material responses: _____

   _____

   _____

   _____

   _____

   G. Specialized positions within the department including hazardous material response teams and equipment, advanced life support teams and equipment, heavy rescue response teams and equipment: _____

   _____

   _____

   _____

   _____

   H. Does the department utilize volunteers to supplement engine companies or other fire department operations? _____

   _____

   _____

   _____

   _____

2. As it relates to your home emergency plan in Exercise 1, answer the following questions.

   A. What is the distance to the nearest fire fighting water source? What does the water source consist of? Include tanks, hydrant systems, rivers, and lakes if applicable.

   _____

   _____

   _____

   B. What is the distance to the nearest fire station? Include station address and phone number.

   _____

   _____

   _____

   C. Is the station a year-round full-time paid department or an on-call volunteer department?

   _____

   _____

   _____

NOTES

1. Smith, Dennis (1978). *History of Firefighting in America: 300 Years of Courage.* New York, NY: Dial Press.
2. National Fire Protection Association (1994). *National Fire Codes.* Quincy, MA: National Fire Protection Association.
3. Riechmann, Jeff. Home Emergency Plan.
4. International City Management Association (1979). *Managing Fire Services.* Washington, D.C.

Chapter 4

Chemistry and Physics of Fire

"There is no fire without some smoke"
John Heywood, 1497–1580

INTRODUCTION

In order to understand and predict fire behavior, a knowledge of the chemistry and physics of fire is necessary. By understanding the physical properties of different materials and how they react to heat and fire, tactics and strategies based on this knowledge can be used to determine the best method of extinguishment. Putting the wet stuff on the red stuff and cooling foundations is no longer the modus operandi in the fire department of today.

Questions

Match the types of fuels to the fire classification in which they belong.

1. Electrically charged fuse box   A. Class A
2. Titanium   B. Class B
3. Forested area   C. Class C
4. Diesel fuel   D. Class D

5. The greater the percentage of surface area to volume of a mass, the (less / more) heat it will take for the material to reach its ignition temperature. _____

6. As the temperature of a liquid rises, the vapor pressure of the liquid (increases / decreases). _____

7. The lowest temperature at which a liquid gives off an ignitable vapor is called the _____ point.

8. The temperature to which an object must be raised to ignite and sustain burning when the outside heat source is removed is called the _____ temperature.

9. Water removes the _____ and _____ sides of the fire tetrahedron.

10. Dry chemical fire extinguishers remove the _____ side of the fire tetrahedron.

11. What information would be obtained by knowing the specific gravity of a petroleum product that is on fire?
    A. The speed and direction in which it will spread when water is applied to it.
    B. If water will create a violent reaction when applied to it.
    C. If water will sink or float when applied to it.
    D. How much water it will take to extinguish the fire.

12. After extinguishment, a petroleum product's ability to flash back is directly related to the product's
    A. fire classification.
    B. specific gravity.
    C. ignition temperature.
    D. flash point.

13. Which of the following is not one of the four sources of heat?
    A. Mechanical
    B. Nuclear
    C. Thermal
    D. Electrical
    E. Chemical

14. The density of a gas compared to an equal volume of air gives what information?
    A. Vapor gravity
    B. Vapor pressure
    C. Vapor solubility
    D. Vapor density

15. When dealing with flammable gases, a visible vapor cloud shows the boundary of the release.
    True   False

16. An oxidizer is a gas that will support combustion, however, may not be flammable itself.
    True   False

17. Molecules in a liquid are vibrating faster than the molecules in a gas.
    True   False

18. The physical state of a material is dependent upon its temperature.
    True   False

Exercises

Exercise 1

In Figure 4.12 of the textbook, if the tank cars are leaking flammable vapors, which would you expect to represent the greatest hazard and why?

_____

_____

_____

Exercise 2

Complete the information regarding the three phases of fire.

PHASE	OXYGEN CONTENT	TEMPERATURE
1. Incipient	_____	_____
2. Free-burning	_____	_____
3. Smoldering	_____	_____

Exercise 3

Identify the four methods of heat transfer.

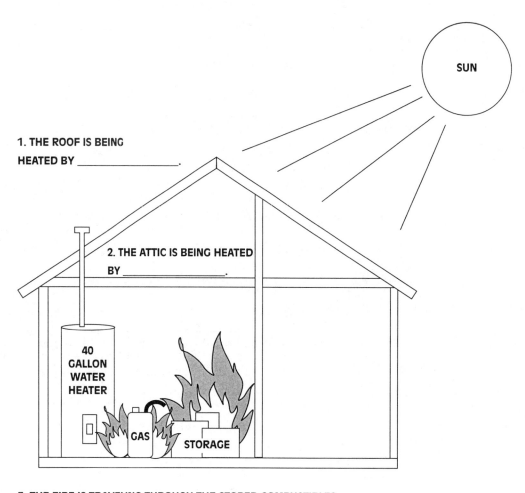

1. THE ROOF IS BEING HEATED BY _____.

2. THE ATTIC IS BEING HEATED BY _____.

3. THE FIRE IS TRAVELING THROUGH THE STORED COMBUSTIBLES BY _____ AND _____.

Figure 4.1

Exercise 4

Read the incident account and answer the questions that follow it.

On March 28, 1994 at 7:36 P.M., the New York City Fire Department received a telephone report of heavy smoke and sparks coming from a chimney at 62 Watts St., Manhattan. The initial response was three engines, two ladders, and a battalion chief. On arrival, they saw the smoke from the chimney but no other signs of fire. The engine companies were assigned to ventilate the roof above the stairs by opening the scuttle and skylight and two three-person hose teams advanced lines through the main entrance to the first- and second-floor apartment doors.

The first-floor hose team forced the apartment door and reported:

- a momentary rush of air into the apartment, followed by
- a warm (but not hot) exhaust, followed by
- a large flame issuing from the upper part of the door and extending up the stairway.

The first-floor team was able to duck down under the flame and retreat down the stairs, but the three men at the second-floor level were engulfed by the flame that now filled the stairway. An amateur video was being taken from across the street and became an important source of information when later reviewed by the fire department. This showed the flame filling the stairway and venting out the open scuttle and skylight, extending well above the roof of the building. Further, the video showed that the flame persisted at least 6½ minutes (the tape had several pauses of unknown duration, but there was 6½ minutes of tape showing the flame).

Damage to the apartment of origin was limited to the living room, kitchen, and hall. Closed doors prevented fire spread to the bedroom, bath, office, and closets. There was no fire extension to the other apartments and no structural damage. The wired glass in the skylight was melted in long "icicles" and the wooden stairs were mostly consumed.[1]

1. What is the fire fighting term used to describe the explosive nature of this type of incident?

2. In which phase of fire was the fire in the first-floor apartment burning immediately prior to the door being forced open by the first-floor hose team?

3. Which side of the fire tetrahedron was not present prior to the door being forced open?

4. Why did the introduction of the missing element cause such an intense reaction?

5. Which phase of burning did the fire enter immediately after the door was forced open?

6. On the fireground, what are three visual indicators that conditions exist for an incident of this type?
    1.
    2.
    3.

Exercise 5

Read the incident account and answer the questions that follow it.

On March 8, 1989 at 8:01 A.M., the Oklahoma City Fire Department was dispatched to a structure fire. Upon arrival, they found a one-story frame residence heavily charged with smoke. The fire had vented itself with fire and smoke coming through the roof. By 8:08:22, an interior attack was in progress. Three firefighters had advanced a 1¾ inch hose line in the front door through a thirty foot front room to a bedroom where the seat of the fire appeared to be. At 8:09:24, it is speculated that the three firefighters had knocked down the visible fire in the bedroom, laid down the hose line, and were exiting the building.

At that time, there was a sudden increase in the intensity of the fire. It engulfed the front room, fatally injuring the three firefighters. Another firefighter who had entered the structure and attempted to open a window narrowly escaped out the front door as the fire came toward him. Immediately prior to the increase in intensity of the fire, this firefighter had observed heavy smoke banking down from the ceiling; however, from his knees he could see the back of the room under the smoke. He had noticed intense heat as he stood to try and raise the window. He had immediately gone back to his knees because his ears were burning. At that point, he looked back and saw fire rolling toward him. After exiting the structure, he immediately reentered with a hose line. Fire activity was still intense. He could knock it down with the hose line but it kept rolling back. At that point, there was no visibility in the structure, making it extremely difficult to locate the firefighters who were down.

Later, it was determined that the structure had been added on to twice. This had resulted in hidden, enclosed spaces where the existing and new roof lines overlapped. Although it appeared the fire had vented itself and been extinguished, there were fire and hot gasses in the concealed spaces that had not been ventilated. As the ceiling began to fall, this allowed the heat and fire gases from those areas to escape into the structure.[2]

1. What is the fire fighting term used to describe the sudden increase in intensity at this incident?

2. Which phase is a fire of this type burning in?

3. On the fireground what conditions would present the opportunity for an incident of this type?

4. What are the indicators that this type of incident is imminent?

5. What actions are indicated when the above indicators are present?

_____

_____

_____

Assignments

Having an understanding of the properties of flammable liquids and vapors is necessary, not only for fire fighting, but important for general public safety. Incidents involving people who did not have this understanding include the following examples:

An individual came home to the smell of gas in his home, reached over and flipped on the light to see where it was coming from. The electrical spark from the light switch ignited the flammable vapor blowing out all the windows in the home and blowing the individual out the door.

In another instance, two people were working on a project for which they needed some propane. Propane is a gas at ambient temperatures and pressures. It is compressed into a liquid and stored in containers under pressure for transportation and storage. The people proceeded to the propane tank with a bucket and opened the valve expecting to fill their bucket with liquid propane. They were taken by surprise as they suddenly found themselves standing in a freezing cloud of propane. Luckily, there was no ignition source nearby.

A mistake made by an individual involved a natural gas leak with fire coming from a broken pipe in the ground. The first inclination of most people is to extinguish a fire, which is just what this person did. Extinguishing the fire without controlling the leak allowed a vapor cloud to form. When the vapor cloud found an ignition source, it ignited. No one was hurt and there was no additional attempt to extinguish the fire until the leak was controlled.

A material is considered a liquid when at normal temperatures and pressures it has the characteristics of a liquid. The three major classifications of liquids that have ignitable vapors are: flammable, combustible, and pyrophoric. Flammable liquids have a flash point below 100°F (38°C). Combustible liquids have a flash point above 100°F (38°C) and below 200°F (93°C). Pyrophoric liquids ignite at or below 130°F (54°C) on contact with air.[3]

Knowing the physical characteristics of these liquids, such as the concentrations of vapors that will burn, which is specified by the upper flammable limit (UFL) and the lower flammable limit (LFL), the specific gravity, vapor density, flash point, and ignition temperature give us valuable information regarding the safest and most efficient means of control during an incident involving them.[4]

Assignment 1

Complete the chart and answer the questions that follow using the information from it.

Product	Flash Point Fahrenheit	Ignition Temperature Fahrenheit	Flammable Limits LEL UEL	Specific Gravity	Vapor Density
Gasoline	–36 to –45	536 to 853	1.4 to 7.6	0.8	3 to 4
Kerosene					
Diesel					
Alcohol					
Propane					

1. Which product has the greatest volatility?

   _____

2. Would water be an effective extinguishing agent for gasoline? _____ On what did you base your answer?

   _____

   _____

3. Which products have vapors that will stay close to the ground and seek out low spots?

   _____

   _____

4. On a winter day with temperatures around 40°F, which would present the greatest hazard, a diesel spill or a gasoline spill?
   A. Diesel
   B. Gasoline
   C. Both would be equally hazardous

5. Which classification of liquids do the following belong in?

   _____ A. Gasoline   1. Flammable
   _____ B. Diesel   2. Combustible
   _____ C. Kerosene

Assignment 2

For those of you who have not spent much time in the kitchen, there are some simple steps you can do there that illustrate things like specific gravity, solubility, and reactivity. Record your observations of each of the steps that follow and answer the questions as they relate to each step. Include information, when apparent, on specific gravity, solubility, and reactivity.

1. Step 1 Fill a clear glass half full of water. Add three to four drops of cooking oil.

   _____

   _____

   _____

   Step 2 Stir the oil and water mixing well.

   _____

   _____

   _____

2. Step 1 Add a pinch of baking powder to 1 tablespoon of water.

   _____

   _____

   _____

NOTES

1. Bukowski, Richard W. *Modeling a Backdraft Incident: The 62 Watts Street (NY) Fire.* Gaithersburg, MD: NIST, Building and Fire Research Laboratory. Published in *NFPA Journal, 89* (6) Nov/Dec 1995.
2. American Heat (September, 1989). *Fatal Flashover: Oklahoma City, OK*, St Louis, MO: American Heat Video Productions, Inc., Volume 4, Program 4.
3. IFSTA (March, 1988). *Hazardous Materials for First Responders.* Stillwater, OK: Fire Protection Publications, Oklahoma State University.
4. Isman, Warren & Carlson, Gene (1980). *Hazardous Materials.* Encino, CA: Glencoe Publishing Co., Inc.

Chapter 5

Public and Private Support Organizations

"Men will find that they can prepare with mutual aid far more easily what they need, and avoid far more easily the perils which beset them on all sides, by united forces."
                                Benedict Spinoza, 1632–1677

INTRODUCTION

The fire fighting force of today is a highly trained organization. Even so, it is a wise firefighter who knows where to turn for additional information and support when necessary. Doing your homework before an incident and knowing what types of organizations are available and what each can contribute makes a substantial difference in how an incident will progress.

Support organizations can be divided into several categories. These include pre-incident support, incident support, and post incident support. The types and numbers of agencies that are utilized depend upon the size and scope of the incident.

For a structure fire in a single family residence, pre-incident support would include technical and manipulative training in structural fire fighting. Incident support could consist of local emergency medical assistance and transport of victims in the event of injuries, local law enforcement for street control, and response from the arson unit if the origin of the fire is of a suspicious nature. Post incident support could include the local utility company, food and lodging assistance for the family from the Salvation Army, local churches, or volunteer groups.

As the size and scope of an incident increases, support needs are also increased. The National Fire Prevention Association Standard 1201-2-6 on Disaster Planning states: "Comprehensive response plans shall be prepared in writing describing the fire department role and providing for management and coordination of all public and private services called into action in natural and technological (man-made) disasters."[1]

In disasters involving large scale destruction too great to be handled with local resources, the channels are opened to mobilize the necessary support. Local officials contact the State Office of Emergency Services (OES) for assistance. OES then contacts the Federal Emergency Management Agency (FEMA). FEMA personnel assess the situation to determine if it meets federal criteria to be declared a disaster area. If it does meet the criteria, the President is contacted and advised of the situation. The President then declares the disaster. Local private and volunteer organizations including the Salvation Army, Red Cross, and the Volunteer Organizations Active in Disaster (VOAD) work to offer immediate food and shelter to victims. FEMA mobilizes their Disaster Response team and other federal agencies that will be needed. These could include the Department of Transportation, Army Corps of Engineers, Department of Agriculture, Federal Military Personnel, and many others. These private and public support organizations work to support the efforts of local personnel. FEMA continues support after the incident, participating in the long-term recovery process of the area.

Many communities stage disaster drills involving local public and private support organizations. Training together and developing good communication between different agencies before an incident builds a solid foundation that can be built upon when an incident occurs.

Questions

For each of the numbered items, list the letter of the agency that could provide the needed assistance.

_____ 1. A copy of the *Fireline Handbook*.

_____ 2. Information on fire apparatus standards.

_____ 3. Training for emergency responders in infectious disease control.

_____ 4. A copy of the *Emergency Response Guidebook*.

_____ 5. Information on summer fire fighting positions.

A. American Red Cross
B. U. S. Department of Agriculture
C. Department of Transportation
D. National Fire Protection Association
E. National Wildfire Coordinating Group

6. Urban Search and Rescue (USAR) teams are considered to be task forces.
   True   False

7. The United States Fire Administration (USFA) was established in 1974 by the Federal Fire Prevention and Control Act.
   True   False

8. The Federal Emergency Management Agency (FEMA) has Type 1 Incident Management teams made up of personnel from many agencies available to respond to all types of disasters.
   True   False

9. Which agency assigns a classification based on risk, which is sometimes used by fire insurance companies in determining insurance rates?
   A. Insurance Services Office
   B. Fire Marshal's Association of North America
   C. National Fire Prevention Association
   D. Insurance Committee for Arson Control

10. Which of the following would coordinate the allocation of fire fighting resources on large wildland fire incidents?
    A. Environmental Protection Agency
    B. National Interagency Fire Center
    C. Department of Interior
    D. Department of Agriculture

11. Which of the following would coordinate large hazardous material incidents in most areas of the country?
    A. CHEMTREC
    B. Federal Emergency Management Agency
    C. Environmental Protection Agency
    D. National Institute for Occupational Safety and Health

12. When approaching the scene of a vehicle accident that involves a placarded material, the number on the placard can be referenced which of the following ways.
    A. Through the Occupational Safety and Health Administration.
    B. Through the Environmental Protection Agency.
    C. By looking it up in the Emergency Response Guide.
    D. By looking it up in the Fireline Handbook.

Exercises

Exercise 1

Read the information on the incident and complete the questions that follow it.

In Oklahoma City on April 19, 1995 at 9:02 A.M., a bomb exploded 20 feet away from the Alfred P. Murrah Federal Building. The explosion caused part of the nine-story structure to collapse, 10 additional buildings to collapse, serious damage to 25 buildings, facial damage to 312 buildings, and ignited more than 40 cars.[2] When Oklahoma City Fire Department operations at the incident came to an end on May 4, 1995, 169 people, including 19 children were dead, and 601 people had been injured.[3]

As the networks broadcast the incomprehensible chaos, we watched the dedication and professionalism of the men and women who gave their abilities, strength, and hearts to the rescue and recovery operations at this incident. Having prepared and well-trained public and private support organizations contributed immensely to operations during this incident. This included treating the many victims of the incident with dignity and respect.

Oklahoma City Fire Department (OCFD) engine companies began responding to the area of the explosion before they were dispatched. A general alarm was dispatched soon after the first arriving chief officers and engine companies began to assess the devastation. The general alarm brought almost 200 OCFD personnel to the scene very quickly. Mutual Aid agreements with other fire departments brought additional support personnel and equipment to cover areas of Oklahoma City whose fire stations were now empty and to provide additional support at the incident. Off duty OCFD personnel were called back to duty to assist at the incident and with station coverage. By the end of the incident, 26 fire departments had responded to assist at the incident.

The year before, personnel from Oklahoma City agencies, utility companies, and volunteer agencies had attended a training course in large scale disasters at the Emergency Management Institute in Emmitsburg, Maryland. The course, which included training in the incident command system, contributed to the success of a coordinated interagency response and resulted in personnel from different agencies who were familiar with the incident command system and structure. The Oklahoma City Fire Department retained Incident Command and utilized the support that came from numerous agencies, organizations, and volunteers.

The gas and electric companies became involved immediately, working to shut down gas lines and electrical services to the large area affected by the blast, and then to reinstate power for incident operations. Local, state, and federal law enforcement agencies responded to the incident which was considered a crime scene because of the nature of the incident. Law enforcement personnel were involved in securing perimeters, street control, investigating the possibility of additional bombs, and checking the entire building for evidence.[2] Medical personnel responded from local hospitals and the local ambulance service provided medical assistance and transport.

Early in the incident, working through the state Office of Emergency Services, a disaster response was initiated and the Federal Emergency Management Agency was contacted. FEMA sent a group of individuals who respond to declared disasters to support management efforts at the incident. FEMA also activated 11 Urban Search and Rescue (USAR) task forces. Each task force consisted of 56 members and was managed by a USAR Incident Support team which is part of the USAR activation. These teams offered their technical training, skills, and specialized equipment to be utilized on the incident by the incident command staff. The California Office of Emergency Services also responded with valuable support and resources.[4] Other agencies included, but were not limited to: state and federal military personnel for support; the Red Cross who helped with family notification and support; counselors involved in critical incident stress debriefings; public works and street clean-up crews; the city finance officer who was essential in purchasing necessary supplies; electricians; welders; and heavy equipment operators. The telephone company provided cellular phone access to assist in communications. The coroner and State Medical Examiner's Office worked on identification of victims. Building plans and information were obtained from the architect and building contractor. A Multiagency Command Center brought representatives of each agency together to coordinate efforts. Additionally, it is estimated that there were ten thousand volunteers who participated in the many aspects of this incident.[2]

1. List three private and/or public agencies that would apply in each category: pre-incident support, incident support, and post incident support and what types of skills or service they provided.

    PRE-INCIDENT SUPPORT

    1. _____

    2. _____

    3. _____

    INCIDENT SUPPORT

    1. _____

    2. _____

    3. _____

    POST INCIDENT SUPPORT

    1. _____

    2. _____

    3. _____

Exercise 2

Read the scenario and answer the questions that follow it.

A tractor trailer rig with two tank cars of liquefied propane lost control as it exited a busy section of state highway. A rescue response that included two three-person engine companies and a battalion chief was dispatched immediately. Upon arrival, the first-in engine company found the rig on an uphill highway exit with the rear tank over on its side. The tank was intact with no leaks; however, in that position, the relief valve was under the level of the liquid. It was determined after talking to the driver that the tank was full. The fire department's five-member hazardous material response team was dispatched to the incident because of the potential for release of the hazardous material.

The incident, which began early in the day, took several hours to resolve as temperatures reached the 100 degree mark. Firefighters stood ready in full personal protective equipment with safety lines as some of the product was off loaded. Because the tank was on its side the discharge valve was in a position that allowed only about half of the product to be off loaded. Two heavy wreckers were brought in to right the tank, after which the remaining product was off loaded. Traffic on the state highway was suspended during operations involving product transfer and repositioning of the tank. The highway exit was closed for the duration of the incident.

1. List two private and/or public agencies other than fire department resources that would have been beneficial in preparing for an incident of this type.

    PRE-INCIDENT SUPPORT

    1. _____

    2. _____

2. List two private and/or public agencies other than fire department resources that would have been utilized during this incident.

    INCIDENT SUPPORT

    1. _____

    2. _____

Assignments

1. Which fire code and edition is in use in your area?

    _____

2. Which local agencies in your area provide temporary assistance to people who lose their homes to fire?

    _____

Look up liquefied petroleum gas (LPG) in the U.S. Department of Transportation's *Emergency Response Guidebook* and answer the following questions.

3. What is the ID number for LPG?

4. What is the Guide Number for LPG?

5. List two potential LPG fire or explosion hazards listed in the Guide.

6. List two health hazards associated with LPG listed in the Guide.

7. What is the recommended action to be taken for an LPG spill or leak without fire?

NOTES

1. National Fire Protection Association, *National Fire Codes*. Quincy, MA: National Fire Protection Association.
2. Marrs, Gary (September, 1995). *Report from Fire Chief Gary Marrs*. Saddle Brook, NJ: Fire Engineering.
3. Davis, Gary (September, 1995). *Victims by the Hundreds: EMS Response and Command*. Saddle Brook, NJ: Fire Engineering.
4. Ghilarducci, Mark (November, 1995). *USAR IST: Providing Search and Rescue Support in Oklahoma City*. Saddle Brook, NJ: Fire Engineering.

Chapter

6

Fire Department Resources

"Resource—Any supply that will meet a need"
Thorndike & Barnhart

INTRODUCTION

The people in a fire department are the department's most valuable resource. From the bucket brigades of yesterday to the advanced equipment used today, the firefighters are the ones who have been and will continue to be responsible for the protection of life and property. The position of firefighter requires ingenuity, adaptability, compassion, and the ability to think quickly on your feet. It also requires interpersonal relationship skills that are necessary to live and work with your company and interact with the public on a regular basis, as well as under extremely stressful circumstances.

The National Fire Protection Association has developed the National Fire Code with specific standards listing minimum requirements for all aspects of the fire service. These standards include requirements for fire department resources such as personnel, personnel protective equipment, and fire apparatus. An excerpt from NFPA Standard 1201 on Fire Company Procedures and Staffing states, "Personnel designated to respond to fires and other emergencies shall be organized into company units or response teams and shall have appropriate apparatus and equipment assigned to such companies or teams."

Fire department resources and operating procedures continue to be developed and modified as additional needs are recognized and newer technologies become available.

Questions

1. Having an automotive repair facility is advantageous for a fire department because of the specialized knowledge of fire pumps and hoists capable of handling the weight of fire apparatus that are needed.
   True   False

2. Cotton jacket rubber lined hose must be completely dried before reloading on fire apparatus to avoid mildew of the jacket and a build up of acid in the liner.
   True   False

3. Synthetic hose is heavier than cotton jacket rubber lined hose; however, it may be drained and loaded back on the truck wet.
   True   False

4. Rescue tools can exert thousands of pounds of force and cannot tell the difference between metal, flesh, and bone.
   True   False

Match the attribute to the pump it belongs to.

_____ 5. Self priming.

_____ 6. Can take advantage of pressure coming in from suction side.

_____ 7. Can draw water from a static source.

_____ 8. Is not jammed by small amounts of debris.

A. Centrifugal pump

B. Positive displacement pump

_____ 9. Discharges a specific amount of water each time it cycles.

_____ 10. Can build up large amounts of pressure without discharging any water.

11. What type of pump is most commonly used as the main pump on fire apparatus?
    A. Positive displacement pump   C. Centrifugal pump
    B. Volume pump   D. Pressure pump

12. Which three attributes constitute a triple combination pumper?
    A. Water tank, pump, hose   C. Ladders, hose, rescue tools
    B. Rescue, fire, medical equipment D. Radio, warning lights, siren

13. What is the purpose of a relief valve on a centrifugal pump?
    A. It allows water to circulate in the pump to keep it cool while no water is flowing.
    B. It allows water to be redirected from the discharge side of the pump back to the suction side to avoid a pressure surge at the nozzle.
    C. It is used to evacuate the air out of the pump.
    D. It is used to protect the main suction valve from damage when there is debris in the water source.

14. Which of the following describes a straight wye.
    A. A single male coupling on the intake side with two female couplings on the discharge side.
    B. Two female couplings on the intake side with a single male coupling on the discharge side.
    C. A female coupling on the intake side, a male coupling on the discharge side and a one-inch male coupling in the center.
    D. A single female coupling on the intake side with two male couplings on the discharge side.

15. What is the fitting called that would connect a 1½ inch National Standard male coupling on a gated wye to a 1 inch Iron Pipe female coupling on a hose?
    A. An adapter   C. A reducer adapter
    B. A reducer   D. An increaser adapter

16. Which of the following is a power fan **not** used for?
    A. To protect exposures from radiant heat
    B. To pressurize stairwells
    C. To evacuate smoke from a structure
    D. To reduce the need for vertical ventilation

17. The larger the diameter of the hose the (more / less) pressure is lost due to friction.

18. What are the two phases of fire that can be used for training purposes in a burn building?
    1. _____
    2. _____

Exercises

Exercise 1

Identify the fittings in Figure 6.1.

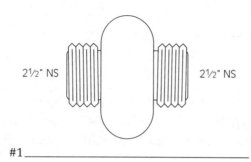

#1 _____

_____

#2 _____

_____

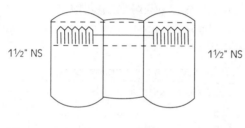

#3 _____

_____

#4 _____

_____

#5 _____

_____

#6 _____

_____

Figure 6.1

Exercise 2

Identify the hand tools in Figure 6.2.

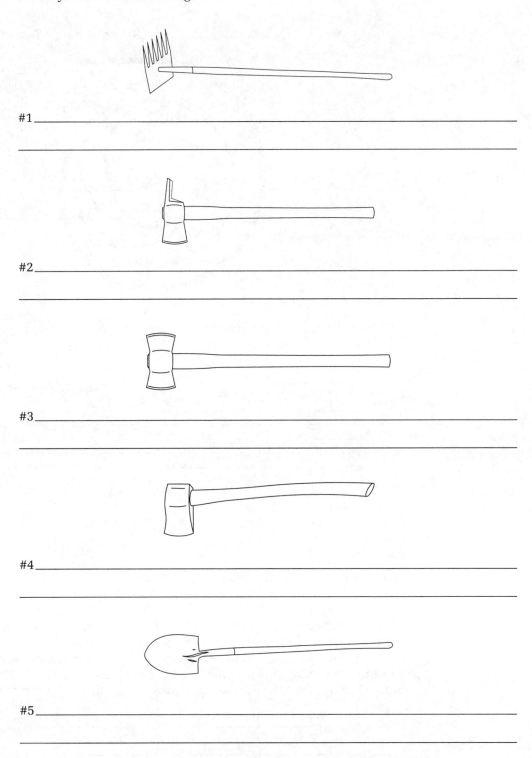

#1 _____

_____

#2 _____

_____

#3 _____

_____

#4 _____

_____

#5 _____

_____

Figure 6.2

Exercise 3

Identify the parts of the centrifugal pump in Figure 6.3.

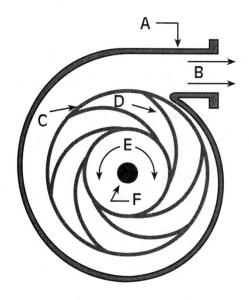

A. _____

B. _____

C. _____

D. _____

E. _____

F. _____

Figure 6.3

Exercise 4

Read the information on the following scenarios and answer the questions that follow.

A fire of unknown origin has been reported midmorning in the lightly forested area above the town of River View. There are dirt roads leading into the area of origin, however, much of the area that is now burning is inaccessible by vehicle. The fire is spreading moderately upslope through fuels of grass and duff toward areas of heavier brush. Upslope from the fire are some narrow canyons leading to a ridge top.

1. What types of fire apparatus and heavy equipment could be used on a fire of this type?

   _____
   _____
   _____

2. Which types of hand tools could be used by fire crews and what would each be used for?

3. What kinds of air support could be used on this fire and on which areas of the fire would each be utilized?

4. If a stream was dammed up to form a small pond and a portable pump was used in the pond for water supply, what kind of pump would be used and why?

Exercise 5

A report of smoke in a single story residential structure was reported at 21:30. Three three-person engine companies and a battalion chief were dispatched to the incident. Upon arrival, engine companies found a structure lightly charged with smoke. While one company made entrance to the attic, another worked outside to try and determine the fire location. The attic was clear of smoke and the fire was found to be in a water heater closet attached to the back of the structure. The fire had spread to the ceiling above the closet and into the bedroom wall of the structure. A ventilation crew was utilized on the roof while the interior attack crew opened up the wall in the bedroom. Salvage operations were also performed by the interior attack group.

1. What personal protective equipment would be worn at a fire of this type?

2. What sizes and types of ladders would be used at this incident?

3. What types of equipment could be used for ventilating the roof?

   _____

   _____

4. What type of equipment could be used for pulling the interior wall and the salvage operations?

   _____

   _____

Assignments

The type of apparatus and equipment selected by a fire department, in addition to the NFPA requirements, depends on a variety of factors. Some of the factors that influence the selection are the geography, types of industries, population density, and weather patterns found within the department's jurisdiction. Other factors include the different types and levels of services provided by the department.

Contact the local fire department or fire department in the jurisdiction of your choice to find out the following information about their apparatus and equipment.

Assignment 1

1. Types of apparatus, engines, patrols, squads: _____

   _____

   _____

   _____

2. Type of fuel used in apparatus: _____

   _____

3. Terrain capabilities: _____

   _____

4. Water tank capacity: _____

   _____

5. Pump type and GPM capacity: _____

   _____

6. Hose:

Hose	Material	Diameter	Number of Feet Carried
Quick attack lines			
Supply lines			
Hard suction hose			
Hard rubber lines			
Other			

7. Does the apparatus have a fixed or portable heavy appliance? If yes, what is its maximum flow in gallons per minute and how many minutes will it take to empty the tank at maximum flow?

   _____

8. What types of fittings are necessary to hook up a supply line to the area's fire hydrants?

   _____

   _____

9. What medical aid equipment is carried on the apparatus?

   _____

   _____

10. What rescue equipment is carried on the apparatus?

    _____

    _____

11. What tools are carried for ventilation?

    _____

    _____

12. What tools are carried for mop-up operations?

    _____

    _____

13. Are there tools carried for wildland fire fighting? If so, what are they?

    _____

    _____

Assignment 2

1. What additional equipment is carried on the apparatus as a result of local needs?

    _____

    _____

    _____

    _____

Assignment 3

1. Is the local fire department's headquarters, or the fire department's headquarters in the jurisdiction of your choice, located at a fire station or in a separate location?

    _____

2. Does the department have its own automotive shop? If not, where are the major repairs done?

    _____

    _____

Chapter 7

Fire Department Administration

"Leadership and learning are indispensable to each other."
John F. Kennedy

INTRODUCTION

Fire department administration is a multi-level process with each person in the department a section of the integral chain that binds it together. The Mission Statement of a fire department provides the overall goals. Objectives are then developed to accomplish the goals.

The fire department maintains a semi-military structure from staff positions through line positions. With this structure, a unity of command and a chain of command is established. In line positions, this allows the rapid mobilization of personnel and equipment and the ability to determine and implement tactics and strategies for effectively controlling an incident of minor or major size.

Questions

1. Unity of command is when the local governing body and the command staff agree with the chief on fire department goals.
   True   False

2. Principles of command were designed specifically for use by the command staff in the administration offices of a fire department.
   True   False

3. Successful delegation of authority requires competent and well-trained personnel.
   True   False

4. The one department concept encourages non-standardization as much as possible.
   True   False

5. Which of the following statements is **not** true regarding the chain of command?
   A. It allows those in command to maintain control.
   B. It is the formal path for communication.
   C. It is used for line functions not staff functions.
   D. It allows for transfer of authority on incidents.

6. In fire fighting, what is the accepted range for the number of people a supervisor can effectively supervise?
   A. One to three   C. Two to four
   B. Three to seven   D. Four to six

Match the characteristic to the type of fire department it belongs to.

_____ 7. Volunteer Fire Department
_____ 8. Combination Fire Department
_____ 9. Public Safety Department
_____ 10. Paid Fire Department

A. 90% of department funding can go for personnel costs.
B. Represents 90% of the fire departments in the United States.
C. Combines law enforcement and fire fighting under the same department.
D. Utilizes paid personnel and volunteers or reserves.

11. Which of the following is **not** one of the four principles that lay the groundwork for incident effectiveness?
    A. Company drills
    B. Equipment maintenance
    C. Division of labor
    D. Station programs
    E. Station maintenance

12. Which of the following is considered the best method of communication when transferring command at an incident?
    A. Radio or telephone
    B. Face to face
    C. Electronic
    D. Written

Exercises

Exercise 1

Complete the organizational chart using the chain of command for paid fire departments described in the textbook.

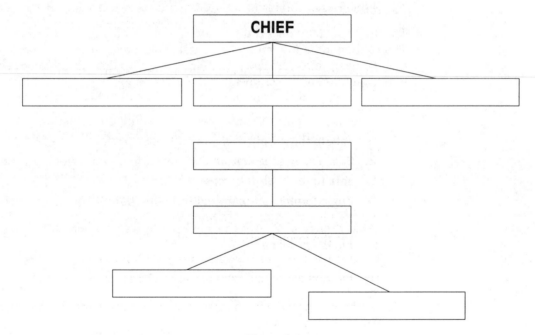

Figure 7.1

In the following exercises read the information given and complete the questions that follow using the organizational chart from Exercise 1.

Exercise 2

The chief of the fire department has issued a directive concerning his desire for increased customer service to the public served by his department. His goal is to make all phases of fire department operations more user friendly including permitting and inspection processes.

How would this directive reach all ranks in the department?

_____
_____
_____
_____

Exercise 3

The driver operator on Y Shift is having a problem with the driver operator on Z Shift. The equipment is low on fuel and water on a regular basis when Y Shift comes on duty to make shift change with Z Shift. Driver operator Y has repeatedly brought this to the attention of driver operator Z. Driver operator Z does not think it is a big deal and points out to driver operator Y that he has all of his shift to top everything off.

1. What is firefighter Y's next step in handling this situation?
   A. Talk to the Z Shift company officer.
   B. Start leaving the equipment low on fuel and water at shift change.
   C. Talk to the Y Shift company officer.
   D. File a formal complaint with the fire chief.

2. After firefighter Y takes the appropriate action, what path would be followed to resolve the problem?

_____
_____
_____

Exercise 4

A firefighter has been asked by a friend who is a teacher to participate in a fire prevention program at a school. The firefighter will be on duty at the time the program is scheduled and the school is not in the response area of the station. The firefighter will submit a memo notifying the proper people of the program and requesting permission to attend. How will this memo be routed?

_____
_____
_____
_____

Assignments

1. Make an organization chart showing the chain of command in the local fire department or fire department in the jurisdiction of your choice.

2. Work through the steps of the management cycle in regard to your career goal. Start with a brief mission statement. Next, write a brief statement for each of the components in the management cycle. In the last step, evaluate the progress you are making toward your goal and objective.

   Mission Statement: _____
   _____
   _____

   Planning: _____
   _____
   _____

   Organizing: _____
   _____
   _____

   Staffing: _____
   _____
   _____

   Directing: _____
   _____
   _____

   Controlling: _____
   _____
   _____

   Evaluating: _____
   _____
   _____

Chapter 8

Support Functions

INTRODUCTION

The numbers and types of support functions found within a fire department are the network that allows the work to be accomplished in an orderly and efficient manner. Having people with expertise in repairing apparatus and equipment, receiving, assimilating, and disseminating information to the field, researching financial information and providing an appropriate budget are only a few of the necessary support functions that contribute to the order and efficiency of a department.

Questions

1. When an incident increases beyond the immediate capabilities of the dispatch center, expanded dispatch is activated at the emergency operations center.
   True   False

2. Expanded dispatch is usually set up in the same communications center as the regular dispatch center.
   True   False

3. When requests for emergency medical services come into dispatch, first aid can now be started before emergency units arrive on scene.
   True   False

4. Since the fire department is a public service agency, emergency costs cannot be recovered from the individual responsible for the emergency.
   True   False

5. A system analyst maintains the computer system and assists the users with program applications.
   True   False

6. The business manager of a fire department has usually come up through the fire suppression ranks.
   True   False

7. Who tracks the resource orders on an expanded incident?
   A. Incident commander   C. Dispatcher
   B. Operations chief   D. Dispatch recorder

8. Which of the following information is reflected on parcel maps?
   A. Property ownership   C. Longitude and latitude
   B. Topography   D. Major landmarks

9. For which of the following are arson unit personnel **not** used?
   A. Determining the cause of a suspicious fire
   B. All fires of unnatural origin
   C. Fires where a crime may have been committed
   D. Fires where a serious injury or death occurred

Match the support function to the type of support to which it applies.

10. Vehicle maintenance   A. Technical
11. SCBA maintenance   B. Manipulative
12. Investigative
13. Weather service information
14. Chemical properties

15. Enhanced 911 dispatching systems give what information to the dispatcher when the call comes in?

    _____

16. List four areas that make an information systems unit a necessary support function for a fire department.

    1. _____
    2. _____
    3. _____
    4. _____

Exercises

Support functions at a fire may involve anything from a cold drink to setting up a fire camp. The chief officer who shows up with hot coffee in the early morning hours for crews that have worked all night on a structure fire provides a most welcome support function and certainly makes points with the firefighters. At a large wildland incident, fire camp can become the size of a town with the corresponding needs of one. Sometimes fire camps are close enough to town to utilize some of the facilities there. Other times, they are not and everything needed must be brought in for the camp. In order to provide the resources for an incident of this type, planning must be done in advance of the potential fire or disaster. Having contracts in place with private businesses who will be able to supply the necessary items when needed contributes to their timely receipt.

Exercise 1

Figure 8.1 is an example of an Incident Command System (ICS) form, ICS 209, Incident Status Summary at a wildland fire incident.

The 209 takes information from the many specialized support units on the incident and combines it for a summary of the incident. In the initial stages of an incident, it is helpful to update the 209 frequently as more resources are committed to the incident. As the incident progresses, the form is completed daily and forwarded to the agency responsible for coordination of resources in the your area.

Information on the 209 is generated in the Plans Division of the ICS System. It includes specialized information from fire behavior specialists on control problems; meteorologists on expected temperatures, humidities, and weather; fire line status from the situation status unit; and information on the agencies assigned to the incident along with the total number of personnel from the resource status unit. The SR and ST under Kinds of Resources on the ICS 209 stand for Single Resource and Strike Team.

INCIDENT STATUS SUMMARY
(See reverse for general instructions)

1. Date	Time	2. INITIAL ☐ UPDATE XX FINAL ☐	3. Incident Name	4. Incident Number
08/03/96	1800		Borel	CA-SQF-1512

5. Incident Commander	6. Jurisdictions	7. County	8. Type Incident	9. Location	10. Started
Dague	SQF	Kern	Wildland Fire	T27 R32 Sec:4,5,6,7, 8,9,16,17,18	Date 08/01/96 Time 14:00

11. Cause	12. Area Involved	13. % Contained	14. Expected Containment	15. Est. Control	16. Declared Controlled
Human	2650	100%	Date 08/03/96 Time 1800	Date 08/05/96 Time 1800	Date ___ Time ___

17. Current Threat	18. Control Problems
Structures in Black Gulch, Keysville, Rancheria, Arch Sites, and timber.	Steep inaccessable Terrain, erratic winds, hot dry weather.

19. Est. Loss	20. Est. Savings	21. Injuries	Deaths	22. Line Built	23. Line to Build
Unknown	Unknown	0	0	860 Chains	0

24. Current Weather	25. Predicted Weather	26. Costs to Date	27. Est. Total Cost
WS 15-20 Temp 90° WD S/SW RH 16%	WS 15-20 Temp 65° WD E/NE RH 59%	$1,404,881	$2,500,000

28. AGENCIES

29. RESOURCES KIND OF RESOURCE	USFS SR	ST	BLM SR	ST	KRN SR	ST	CDF SR	ST	NPS SR	ST	CCC SR	ST	NWC SR	ST	LAC SR	ST	NWS SR	ST	ORC SR	ST	SBC SR	ST	TOTALS SR	ST
ENGINES	10	3	3			1		2				1											14	6
DOZERS		1		1		1																		3
CREWS	24		1								6												25	6
HELICOPTERS	7		1																				8	
AIRTANKERS	7																						7	
TRUCK COS.																								
RESCUE/MED.																								
WATER TENDERS	6		4		1																		11	
OTHER																								
OVERHEAD PERSONNEL	123		15		24		9		5		2				2		2		1		1		184	
TOTAL PERSONNEL	627		60		39		43		5		92		2		2		2		1		1		874	

30. Cooperating Agencies
California Highway Patrol, Kern County Sheriff, Kern Valley High School, Kern County Search and Rescue

31. Remarks Fire spread was minimal. Fireline construction was completed this afternoon. The mop-up objective of 500 feet has begun on all areas of the line. The incident was declared contained at 1800 hours on 08/03/96. The DMOB Plan has been developed and the release of excess personnel will begin on 08/04/96 at 0600. Recreational use of the Kern River will be opened on 08/04/96.

32. Prepared By	33. Approved By	34. Sent To Date Time By

NFES 1333   9/86 ICS 209

Figure 8.1

The logistics section of ICS is responsible for meeting the needs of the personnel on the incident. List three areas that will need logistical support and coordination on an incident of this size and briefly describe what filling those needs would entail.

1. _____

2. _____

3. _____

List four support functions that would have been involved in preparing the fire equipment and crews to respond to the above incident.

1. _____
2. _____
3. _____
4. _____

Exercise 2

Assessor's maps can be used to determine property ownership for such things as fire reports or fire hazards. Map design and designators depend on the local system in use. In the following exercise, the first three numbers in the assessor's parcel number reflect the number of the book the map is in, the next two numbers denote the page number the map is on, the following number indicates the different areas on the map, and the last two numbers identify the specific parcel on the map. In Figure 8.2, the assessor's parcel number (APN) 348-121-17 corresponds to the parcel on the map with the star.[1]

1. If a hazard reduction complaint was received by the people who live at APN 348-122-18 about the vacant lot to the east of them, which parcel number would you use to look up information for the property owner of the property with the hazard?

   _____

2. If a fire was started on APN 348-121-18 and spread to the parcel to the south of it, which parcel number would you look up to determine the property owner of the lot the fire spread to?

   _____

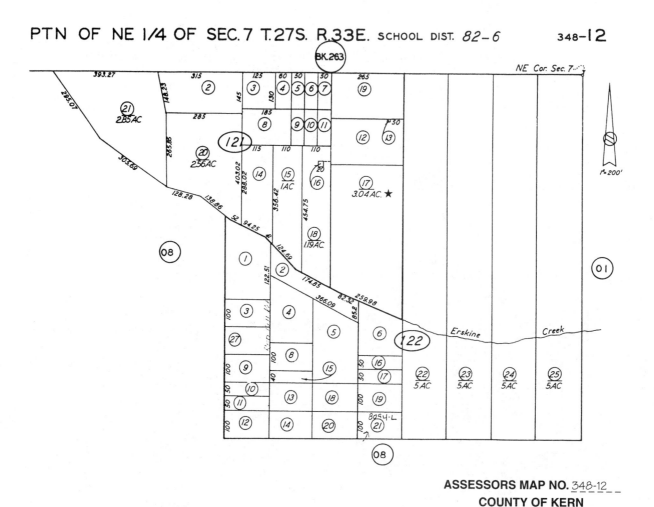

Figure 8.2

Exercise 3

Topographical maps can be used during a wildland fire incident to show personnel the terrain characteristics within and surrounding the fire's boundaries. This information can be used for determining the tactics and strategies that will be used to control the fire, as well as locations for helispots and drop points on the fire. Other uses for topographic maps include use by chaparral management or burn managers to determine boundaries for managed or prescribed burns.

Answer the questions regarding the topographical map in Figure 8.3.[2]

1. Using the ICS 209 in Exercise 1, mark the location of the Borel Fire on the topographical map.

2. Using the parcel map in Exercise 3, mark the area on the topographical map where the parcels are located.

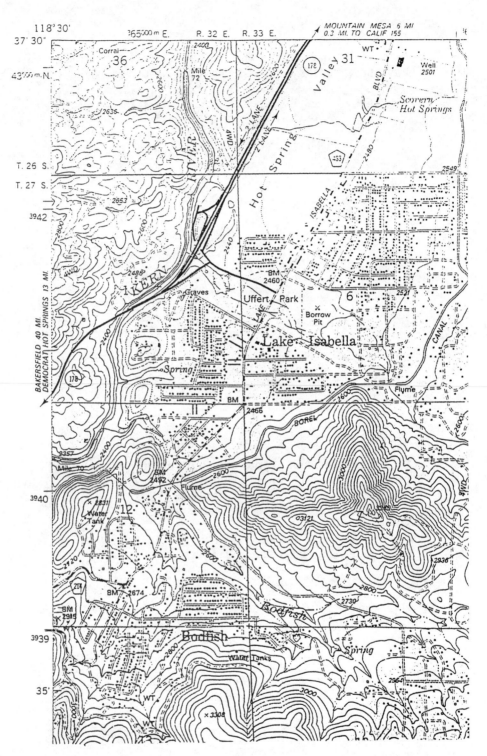

Figure 8.3

Assignments

1. In the department of your choice, what types of support units are available and how many personnel are assigned to the units on a regular basis?

   Dispatch: _____

   Maps / GIS: _____

   Haz Mat Control: _____

   Arson: _____

   Cost Recovery: _____

   Personnel: _____

   Information Systems: _____

   Business Management: _____

   Warehouse: _____

   Radio Shop: _____

   Auto Shop: _____

   _____: _____

   _____: _____

2. Does the department of your choice have written contracts or agreements with private business to help meet increased needs during extended incidents? If so, what types of needs do those businesses fill?

   _____

   _____

   _____

3. Does the department of your choice have organized teams that respond to expanded incidents to fill ICS support functions?

   _____

4. Briefly describe the parcel identification system used in your area.

   _____

   _____

   _____

NOTES

1. County of Kern.
2. U.S. Department of the Interior Geological Survey.

Training

"By training together and using the same operating procedures, safety is strengthened."
Jack Lee, Assistant Fire Management Officer, Sierra National Forest [1]

INTRODUCTION

Training can be defined as educating ourselves in our chosen profession, or developing strength and endurance. Both definitions apply to the profession of firefighter. The position of firefighter requires that our minds and our bodies be well-trained. Extensive technical training is necessary in numerous areas including fire chemistry, tactics and strategies, first aid, hazardous materials, building construction, vehicle construction, codes and ordinances, and extinguishing systems and agents. Manipulative training is necessary to learn and refine techniques used to operate equipment including fire apparatus and pumps, fire hose and fittings, heavy appliances, ladders, rescue tools and equipment, and medical aid equipment. Strength and endurance training is necessary to maintain a physical fitness level commensurate with the physically demanding tasks required of firefighters. Training is designed to offer a reasonably safe environment in which learning and conditioning can take place. Each emergency we are involved in also provides invaluable information as we build upon the experiences of each incident and take that knowledge with us to the next.

As firefighters we must learn from our mistakes. Many of the standard operating procedures (SOP) departments follow today and all of the wildland fire fighting standards are a result of firefighter fatalities. An example of a standard operating procedure pertaining to forcible entry is to wear full personal protective clothing, to approach the door of a structure in a crouched position in front of the wall next to the door carrying a charged hose line evacuated of air, a flash light and pike axe, and to feel the door for heat before attempting entry. If a backdraft condition exists, this positions the firefighter out of the path of the backdraft. While a backdraft condition does not exist at every structure fire, this SOP places firefighters in the safest location for forcible entry to a structure.

An SOP in regard to vehicle fires is to not approach the vehicle from the front or back. Many vehicles have the 40-mile an hour hydraulic bumpers. When the hydraulic fluid is heated to its vaporization temperature, it launches the bumper like a rocket, hurling it at about knee level straight out from the vehicle. While not every bumper at a car fire will do this, not approaching the vehicle in the potential path of the bumper puts firefighters in a safer position for attacking a vehicle fire. Both of the above SOPs were developed and are in place in many departments as a direct result of serious injury or death. There are many standard operating procedures within every fire department. Included in them are SOPs having to do with ladders, hose lines, nozzle pressures and gpms, as well as many others.

Firefighters must be able to relate past training and experience to current incidents. Things do not always go according to plan on the fire ground. If a technique you have been trained in does not work, for whatever reason, firefighters must quickly, often under extreme circumstances, try another way. Being well-trained and combining training, experience, and your department's standard operating procedures allows this to be done without panic or confusion. Firefighters are reminded that they have not caused the situation they are trying to control, but are working to bring it to a close without adding to the problem.

Training builds on previous knowledge. Everyone comes into the fire service with their own base of skills and knowledge. Mechanical skills, medical backgrounds, people skills, and writing skills are all assets upon which additional training will be built.

In fire fighting, training never stops. No matter how long you have been on the job you will be participating in ongoing training. Some parts of this training will be determined by your own department, other parts are mandated by state and federal regulation. It is each firefighter's responsibility to continue to learn from those who have gone before and progress to the level where they are able to train the next generation of firefighters. As a new firefighter, training encompasses the majority of your time. While the experienced members of the crew may be taking a break in the evening, the new firefighter will be, and is expected to be, buried in the books, studying.

Questions

1. Firefighters must be able to relate past training and experience since it is impossible to train for every situation they will encounter during their careers.
   True   False

2. Because of the highly manipulative nature of fire fighting, classroom teaching and learning is rarely used.
   True   False

3. Servicing and cleaning equipment after drills can be relegated to non-safety personnel since these tasks have no training value.
   True   False

4. After initial mastery of a skill, it is necessary to practice to maintain the skill level.
   True   False

5. The first criteria for determining the adequate level of training is
   A. that complex evolutions are completed in the proper sequence.
   B. that the evolution is completed in the required time.
   C. that the evolution is performed in the safest manner possible.
   D. how often the evolution will be performed.

6. When doing a preplan of a high hazard occupancy, which of the following is not an advantage of having a multi-company training drill?
   A. Hose can be used from one engine keeping the rest in service.
   B. Some engine companies will be out of their first-in areas for the duration of the drill.
   C. Hazards can be identified and discussed by first-in engine companies.
   D. Outside areas and buildings can be checked to verify that radio communication capabilities are functional and adequate.

7. Which is an advantage of joint training with other agencies and departments?
   A. Provides information for all involved on each agency's capabilities and equipment compatibilities.
   B. Provides opportunities for training and qualification systems on local, state, and federal scales.
   C. Gives outside agencies an understanding of fire department operational practices and procedures.
   D. All of the above are advantages of training with other departments.

8. The overall purpose of training is
   A. safe operations.   C. co-ordination of forces.
   B. incident effectiveness.   D. swift operations.

9. The most important area of technical and manipulative training is _____.

10. The performance standard of a 0 % acceptable error level regarding safety during training is also referred to as _____ _____.

For each of the following, match the area of training with the nature of the training.

_____ 11. Ladder operations   A. Technical Training
_____ 12. Types of building construction   B. Manipulative Training
_____ 13. Extinguishing agents
_____ 14. Use of ventilation tools
_____ 15. Hose line deployment
_____ 16. Properties of hazardous materials
_____ 17. Completing this workbook

18. List two advantages of timed evolutions.

   1. _____

   _____

   2. _____

   _____

19. What is one of the drawbacks of timed evolutions done in a training environment?

   _____

   _____

Exercises

During training drills, characteristics of equipment and capabilities of personnel become apparent. While the same is true for emergencies, training drills offer a much safer environment in which to determine these things. When training, be aware of all of the information you are receiving. For instance, when raising a ladder you are trained to watch out for overhead lines, to make sure the ladder is at the proper climbing angle with the tip and butt resting evenly, to make sure the dogs are locked and the halyard is tied. Some of the more subtle information you might encounter while performing a ladder drill could be, "I really need to break these gloves in better to get a good grip on the halyard," or "when I lean my head back while raising the fly, the back brim of my helmet hits my SCBA, knocking the front brim down over my eyes." When performing a drill to set up lighting at a night-time fire, you may realize the need to go back and work with familiarizing yourself with the operating instructions on the generator. When these types of information are recognized on a drill, after the drill is complete, don't just forget them. Take the time to resolve or correct them. If you don't, the next time you encounter them may be in an emergency situation and it's guaranteed that at the very least you will be kicking yourself when they present themselves again.

Training drills are usually set up and monitored by training officers. A safety officer is also a standard participant on training exercises. Even so, it is each person's responsibility to be alert to factors affecting their own safety and the safety of those around them. For new firefighters, it is not always easy to distinguish between what is and is not safe, and whether or not their personal protective equipment is functioning properly. Firefighter trainees and probationary firefighters are under pressure to prove themselves capable of doing the job. One firefighter trainee who was involved in her first live structure fire training drill ended up with second degree burns on her upper back even though she had been wearing full personal protective equipment. She was at the front of the hose line with the nozzle; the firefighters behind her were uninjured. When asked what happened she replied, "I didn't really know how much heat to expect. I could feel my back burning, but assumed everyone else was feeling the same thing. I did not realize the extent of my burns until I came out of the structure." When advancing a hose line on a fire, each person can be exposed to different degrees of heat. This can be due to their distance from the fire or having people in front of you that act as a shield. In order to avoid possible injury, be sure to understand the drill and what is expected of you before you begin. No one needs to be injured to prove they can do this job.

A practice to follow during drills is to wear the personal protective equipment (PPE) you will use on the emergency for which you are training. Climbing a ladder and working on a roof in turnout boots and an SCBA is not the same as wearing your uniform work boots and working without the weight of an air bottle or the restricted visibility of a breathing apparatus face shield. Simple manipulative jobs such as starting power fans and chain saws can be hampered by thick gloves.

Learning to keep track of your PPE and other equipment is another important part of training. For example, it is good practice never to lay your gloves or radios down on something. The next thing you know, you will be somewhere else and your equipment will not be with you. One firefighter shares with new recruits the story of the time he ended up going into a structure fire with only one glove on. Immediately prior to entrance, he reached in his pocket and found only the one glove. He ended up fighting the fire while trying to keep the unprotected hand in his pocket as much as possible to avoid getting it burned. It is important to keep track of your PPE equipment, check it at shift change, and after every incident to make sure nothing is missing and everything is where it is supposed to be.

In May of 1986, the National Fire Protection Association adopted NFPA 1403, Standard on Live Fire Training Evolutions in Structures. The Alert Bulletin which follows gives some background information showing the need for this standard and gives an overview of the safety issues stressed in it.[2]

ALERT BULLETIN on Live Fire Training

In the wake of the recent multiple fire fighter fatality training incident in Milford, Mich., the NFPA issues the following alert.

MARTIN F. HENRY
Director
NFPA Public Fire Protection Division

The National Fire Protection Association is conducting a study of the fatal fire fighter training incident which occurred in Milford, Mich., on October 25, 1987. Three fire fighters were killed and three others injured when fire engulfed a two-story wood frame dwelling during a live fire training exercise. (See the incident fact sheet on the facing page.) This incident followed another fatal live fire training incident in Hollandale, Minn., on October 20, in which one fire fighter was killed.

Over the past decade, NFPA records show that 53 fire fighters have died in training accidents involving a range of activities, such as live fire training, smoke drills, physical fitness or hose and ladder evolutions. A primary cause of death in many of the incidents was heart attack. During this period, nine deaths were attributed to live fire or smoke training. These incidents point toward the need for adherence to fire fighter health and safety policies and related standards.

A fatal live training incident in 1982 (see "Two Die in Smoke Training Drill," *Fire Service Today*, August 1982) led the NFPA's Committee on Fire Service Training to develop a standard that would help prevent such tragedies. The committee's efforts led to the adoption, in May 1986, of NFPA 1403, *Standard on Live Fire Training Evolutions in Structures*.

Ongoing training for fire fighters is the cornerstone of good fire protection in today's world. However, the benefits derived from live fire training are negated by the injuries and deaths suffered by fire fighters under unsafe and poorly supervised training conditions. It is therefore essential that all fire departments focus renewed attention on NFPA 1403. Like all NFPA consensus standards,

NFPA 1403 was developed by a committee of experts who represent the nation's fire community. The document was open to public examination and comment, and it is the result of the best thinking available.

Above all, NFPA 1403 stresses safety, and proposes a basic system, one that can be adapted to local conditions to serve as a standard mechanism for live fire training evolutions. The standard deals with the training of fire fighters under live fire conditions and focuses on training for coordinated interior fire suppression operations.

Live fire training in a suitable, acquired building is an accepted means of training fire fighters, because it provides high levels of realism. It can simulate most of the hazards of interior fire fighting at an actual emergency. Evolutions must be planned with great care, and they have to be supervised closely by instructional personnel. NFPA 1403 contains information that is designed to ensure adequate levels of safety while allowing local fire departments some flexibility to use independent judgment based on local situations and the level of training to be accomplished.

The following guidelines for live fire training in acquired buildings are

Note: This bulletin was prepared by the National Fire Protection Association in the public interest and may be reproduced freely with the customary credit to the source. Any questions regarding NFPA 1403 should be directed to Martin F. Henry at NFPA headquarters.

among the several requirements contained in NFPA 1403 (see the standard for complete requirements):

- Inspect the structure for anticipated load bearing capacities.
- Remove hazardous storage and correct hazardous conditions.
- Provide protection for spectators and exposed structures.
- Provide adequate water supply to handle the anticipated fire with a reserve supply equal to an additional 50 percent; provide separate water sources for attack lines and backup lines.
- Provide a pre-burn briefing plan to discuss each evolution and to make specific assignments; all participants must be familiar with the building configuration.
- Do not use flammable liquids.
- Appoint a safety officer to prevent unsafe acts, eliminate unsafe conditions and provide for the safety of all participants. The safety officer shall have no other duties that interfere with these responsibilities.
- Establish a building evacuation plan.
- Provide on-site emergency medical services.
- Designate one person to control the materials being burned and to ignite the fire in the presence of and under the direct supervision of the safety officer.
- Equip all participants with full protective clothing and self-contained breathing apparatus that meets the appropriate NFPA standards.
- Provide qualified instructors, one per five trainees.
- Establish a method of fire ground communications to allow coordination between the incident commander, the interior and exterior sectors, the safety officer, and external requests for assistance.

Exercise 1

Read the following case study and answer the questions that follow.

In Michigan on October 25, 1987 at approximately 8:45 A.M., three firefighters were injured during a structure fire training exercise in a two-story single-family dwelling. The drill was designed to help firefighters recognize the difference between incendiary and non-incendiary fire starts. Fuels, which included combustible furniture, were arranged in several rooms on the first and second floors. The combustible piles were joined together by trails of flammable liquids consisting of gasoline and camp stove fuel.

Firefighters entered the buildings, ignited the first floor fuels and proceeded to the second floor. There was a rapid unanticipated build up of heat resulting in a rapid progression of the fire. The firefighters attempted to escape from the second-story windows. Three firefighters, though injured, did escape. Three others were unable to escape as the structure was quickly engulfed in flames.[3]

1. List four of the safety issues addressed in NFPA 1403 that would decrease the chance of injury in a training incident of this type.

    1. _____
    2. _____
    3. _____
    4. _____

2. What are two areas of training that contributed to survival of three of the firefighters?

    1. _____
    2. _____

Exercise 2

Read the following case study and answer the questions that follow.

On August 20, 1993, what had appeared to be a quiet section of the Glen Allen wildfire suddenly blew up below a crew that was cutting direct line. The crew foreman and four crew members escaped uphill to a horse trail, which was 30 feet above the location where they were working. Two firefighters were unable to escape the flames and died. Two more firefighters were trapped by the fire but were able to protect themselves and survived.

Christopher Barth[4] is one of the firefighters who survived the entrapment. He noted that when the crew left the horse trail to cut direct line, it had appeared that the fire was laying down. He says "there were only a few embers with very little flame." The terrain in which the crew was working included a very steep hillside with loose soil, making movement difficult. The brush below the crew took off, overrunning them within seconds. The word to get out came from the crew members working above Christopher Barth. As he and the crew member with him started moving up the hill, they could see that the other side of the ravine was already on fire and the fire was moving fast. Because of the loose soil and the fire's rapid rate of spread, they realized they would not be to able outrun the fire uphill and began moving farther into the burn instead of up the hill. The fire overran the two firefighters before they could deploy their fire shelters.

When Barth was asked during an interview how he was able to protect himself from the fire, he replied, "Protection is something you think about before you go to the fire, it's not just once you get there. Check your gear every morning to make sure it is working." He went on to say that checking his gear is something he does. Because of that, he had confidence that he was protected. Barth went on to describe the events that followed.

"It sounds like a train. I laid down and tried to protect my face as much as possible, staying in the ground as much as I could. First there is thick smoke, then wind, then the heat. It is quick. I was having trouble breathing and then you are in pain. It moved over my legs, arms then the parts of my face which the shroud was not covering. It probably lasted 15–30 seconds." When asked how he was able to stay put and not panic, Barth says, "It had a lot to do with training, confidence in my gear, and the fact that I was still alive. Rational thought stops for awhile and all you are thinking about is survival. Our training tells us no matter how bad the pain is, don't move. After a time the heat subsides and the flames are gone. You can feel the difference. You stay down because your training and mind tell you not to get up too soon." His advice to firefighters, "Pay attention to your training. Listen to your training and your common sense. Wear your protective gear. If you think your gear is uncomfortable while working on the hill, you should try the bandages in the hospital."

1. What are two areas of technical training that would contribute to your chances of survival in a situation of this type, and how would those areas apply in your decision-making process?

    1. _____

    2. _____

2. What area of manipulative training would contribute to your chances of survival in a situation of this type?

    _____

3. What areas of physical training would contribute to your chances of survival in a situation of this type? (Obviously this is not something you would want to encounter with a hangover of any kind.)

    _____

Assignments

1. In the local fire department or the fire department in the jurisdiction of your choice, how is the training division structured and staffed?

    _____

Is the training division in the department considered a staff or operations function?

2. In the department you selected, what introductory training is required before a new firefighter is allowed to work on a fire crew or engine company?

3. In the department you selected, what type of monthly training is required? Include areas of technical and manipulative training.

4. Is the department involved in offering specialized training and certifications within the department and/or to other agencies? If so, what types of training and certifications?

5. Have you ever been involved in a training drill where you were uncomfortable with the safety precautions being taken? If so, what were the circumstances and was NFPA 1403 being followed?

6. What would be the best action for you to take on a drill when you recognize a safety issue that needs to be addressed?

NOTES

1. Address to the Interagency Wildland Firefighter Safety Workshop 1996, Jack Lee, Asst. Fire Management Officer, Sierra National Forest.
2. ALERT BULLETIN on Live Fire Training, Martin F. Henry, Director NFPA Public Fire Protection Division, Fire Command December 1987.
3. Firefighter Fatality Training Incident FACT SHEET, Thomas J. Klem, Director NFPA Fire Investigations and Applied Research Division, Fire Command December 1987.
4. Fire Shelter Fire Storm Survival, FD, T.V., County of Los Angeles Fire Department, 1994.

Chapter 10

Fire Prevention

"An ounce of prevention is worth a pound of cure."
T.C. Haliburton, 1843

INTRODUCTION

In 1973, the National Commission on Fire Prevention and Control published its report *America Burning*, identifying fire as a national problem. Even with a more than 30% decline in fire death rates since the report, the United States continues to have one of the highest fire death rates in the industrialized world. The United States Fire Administration reports that approximately 5,700 people die and 29,000 are injured by fire each year in the United States. In addition to civilian deaths, 100 firefighters die annually in the line of duty.

Our current methods of fire prevention, which include the areas of engineering, education, and enforcement, have, and continue to be forged with the hammer of lost lives and property. At times, the cost has included entire cities.

Questions

1. Each firefighter must have advanced training and specialized knowledge in the area of fire prevention inspections.
   True   False

2. With hazardous activities where it is impossible to control all fuel and ignition sources, devices and systems can be installed to control or prevent fire spread.
   True   False

3. A violation corrected in the fire inspector's presence need not be documented on the inspection form.
   True   False

4. To have an effective fire prevention program, every violation of the fire code should be aggressively prosecuted.
   True   False

5. Which is **not** necessary when completing a fire prevention inspection?
   A. Reviewing previous inspections
   B. Making an appointment with the owner
   C. Having identification and credentials
   D. Having access to the fire code for reference

6. Which statement most accurately describes the goal of the fire prevention inspection?
   A. Preplan potential fire operations
   B. Locate violations and issue citations
   C. Systematic examination for compliance of codes and ordinances
   D. Educate the owner in fire safety practices

7. Which of the following is **not** true of arson units?
   A. Works closely with law enforcement agencies
   B. Utilized only at fires of suspicious nature
   C. Has knowledge of chemistry and physics of fire
   D. Must follow a set of legal procedures in the collection of evidence

8. Which of the following is **not** one of the three E's of fire prevention?
   A. Education
   C. Effectiveness
   B. Enforcement
   D. Engineering

9. List four avenues or opportunities for presenting the fire prevention message to the community.
   1. _____
   2. _____
   3. _____
   4. _____

10. List two reasons why determination of fire cause is important in terms of fire prevention.
    1. _____
    2. _____

Exercises

REPORTS

Company Inspection and Fire Incident Reports utilize National Fire Protection Association 901, Uniform Coding for Fire Protection. This allows standardized information to be sent into state and federal incident reporting systems. Using this common format, information can then be sent through the Federal Emergency Management Agency to the United States Fire Administration for compilation. This results in valuable information on national fire trends and allows the identification of specific needs at both local and national levels.

COMPANY INSPECTION

The Company Inspection encompasses both the education and enforcement areas of fire prevention. It provides an opportunity to educate owners and employees in fire safe practices by identifying possible hazards and fire code violations. While the goal is voluntary, compliance through education inspections also allows the identification of violations and enforcement of code violations if necessary.

INSPECTION FORM

The fire inspection form contains information about the building and the type of business being done at that location. Figure 10.1 is an example of a fire inspection report form.

**FIRE DEPARTMENT
INSPECTION REPORT**

DATE: 11/05/95

BUSINESS NAME: Penny Wise

GENERAL USE: Thrift Store

OCCUPANCY CLASS: B 2   PROPERTY CLASS: 5 2 1

LOCATION: 2764 Second Street, River View
                Street Address   City

NUMBER OF BLDGS: 01   NO. OF STORIES: 01   BASEMENT (B) ____
                                                                  CELLAR   (C) ____

MANAGER: Bethany McDonald   PHONE: 555-9843

OWNER: Aimee McDonald   PHONE: 555-2525

DISPOSITION

| 1-CORRECTED | 2-WILL CORRECT | 3-VIOLATION NOTICE ISSUED CALL BACK NECESSARY | 4-CITATION ISSUED |

BUILDING	YES/NA	NO/DISP	FIRE PROTECTION EQUIP	YES/NA	NO/DISP
1. EXIT DOORS	/x	/	8. FIRE EXTINGUISHERS	/	2/
2. EXIT SIGNS	/x	/	9. AUTOMATIC SYSTEMS	/x	/
3. EXIT CORRIDORS	x/	/	10. WET STANDPIPES	/x	/
4. AISLE SPACING	x/	/	11. DRY STANDPIPES	/x	/
5. OCCUPANT LOAD SIGN	/x	/	12. ALARM SYSTEMS	/x	/
6. VERTICAL OPENINGS	/	2/	13. FIRE ASSEMBLIES	/x	/
7. WARNING SIGNS	/x	/	14. FIRE WALLS	/x	/
COMMON HAZARDS	YES/NA	NO/DISP	**SPECIAL HAZARDS**	YES/NA	NO/DISP
15. ELECTRICAL	/	1/	22. GREASE HOODS–DUCTS	/x	/
16. FURNACE/BOILER RM	/x	/	23. FLAMMABLE LIQUIDS	/x	/
17. COOKING EQUIPMENT	/x	/	24. LIQUID PROPANE GAS	/x	/
18. HEATING EQUIPMENT	x/	/	25. COMPRESSED GAS	/x	/
19. DECORATIONS	x/	/	26. CHEMICALS	/x	/
20. HOUSEKEEPING	x/	/			
21. WEEDS	x/	/			

REMARKS: #6. Missing and loose ceiling tiles must be replaced or covered.
#8. Fire extinguisher past due for service. Owner will call service company.
#15. Cannot use extension cord for permanent wiring to cash register. Owner corrected in our presence.

CLEARANCE: GRANTED (G) G   INSPECTOR _____
                      DENIED   (D) ____   OWNER/MANAGER _____

Figure 10.1

Exercise 1

Using the information below, complete separate Fire Inspection Reports for the following five businesses. Blank Inspection Report forms are provided at the end of Chapter 10. All of the businesses are one-story, single buildings, B2 occupancies, with no basement or cellars. Property class can be found in *Excerpts From NFPA 901 Uniform Coding* also found at the end of Chapter 10. You are the inspector. The owner will correct the violations unless otherwise noted. No call backs will be necessary.

1. The Sundae Best is an ice cream shop located at 1530 Main Street in River View. It is owned and operated by Maria Gonzales, phone number 555-8243. On inspection, the following safety item was noted: the compressed gas cylinders for the soda fountain need to be secured with a chain. Everything else looks good. Inspection granted.

2. The Curly Q is a beauty shop located at 1530 Main Street in River View. It is owned by Katie McClure, 555-0124 and managed by Irene Johansen, 555-2222. On inspection, the following safety items were noted: there is a hole in the ceiling over the hair dryers that needs to be repaired; the fire extinguisher under the counter needs to be mounted on the wall; an extension cord is being used as permanent wiring to the tanning booth and needs to be removed; an accumulation of weeds and trash outside the back door by the gas shut off valve needs to be removed. Inspection granted.

3. The next building located at 1520 Main Street in River View is vacant. Inspection denied.

4. The River View Market is a retail food store over 930 square meters located at 1500 Main Street in River View. It is owned by Matt Johnson, 555-1234 and managed by George Winn, 555-4321. On inspection, the following safety items were noted: some of the exit corridors in the store and in the back storage area are blocked by merchandise; all fire extinguishers are due for service and fire extinguisher in storage area is missing the service tag; the valve on the OS&Y fire protection system needs to be chained in the open position (valve was chained open in engine company's presence); storage in front of the electrical panel in bakery area must be removed (storage was removed in inspector's presence); storage in back is up to the ceiling and must be removed to 18 inches below ceiling; grease hood in deli area needs to be cleaned. Inspection granted.

5. The Drug Store is a variety store over 930 square meters located at 1510 Main Street in River View. It is owned by Julie Michaels, 555-1246 and managed by Tom Smith, 555-2941. On inspection, the following safety items were noted: fire extinguishers are due for service; sprinkler valve inside the building is inaccessible due to items stored in front of it (storage was removed in inspector's presence); hole in fire wall in bookkeeping area needs to be sealed; door panel is missing on the electrical breaker box in the storage room and needs to be replaced. Inspection granted.

Exercise 2

Conduct a fire prevention inspection of your residence. Complete a Fire Inspection Form noting any hazards or safety items you find. Use a Fire Inspection Report form found at the end of Section 10. If safety violations are found, list the plan of action that will be taken to correct them in the remarks section of the form.

Exercise 3

Incident Reports are completed by the engine company in whose area the emergency originated. Since not all of the incidents that the fire department responds to are fires, there may be separate Incident Reports for different types of incidents. These could include Incident Reports for medical aids, hazardous materials, public services, and false alarms. Some incidents that have fire potential, but no fire, such as vehicle accidents and rescues, may result in only a Medical Aid or Rescue Incident Report. A major spill from a diesel tank with no fire would result in a Hazardous Material Incident Report. A fire with injuries to the public or firefighters would result in a Fire Incident Report and a Medical Incident Report.

Some of the information documented in the Fire Incident Report includes the ignition source, type of material ignited, specific use of the building or property, number of personnel and equipment responding, and type of action taken to extinguish the fire. Figure 10.2 is an example of a Fire Incident Report.

Complete Fire Incident Reports, found at the end of this section, for the following two fires. Use *Excerpts from NFPA 901 Uniform Coding*, found at the end of this section, to obtain the correct numbers for the report.

1. On October 10, 1995 at 13:28, a three-person engine company from Station #1 was dispatched to a structure fire, incident number 3075, at the one-story single-family residence located at 1259 Maple Street in River View. The fire was one mile from the station and the engine was on scene in two minutes. The fire was located in a storage closet containing a water heater with access from the outside of the structure. The 150 foot, 1½ inch preconnected line was pulled and the fire was extinguished with 25 gallons of water. The fire was controlled at 13:34, with damage confined to the contents in the closet. On inspection, the structure was noted to be of ordinary Type III construction with an asphalt roof. The fire was caused by storage of gasoline in an improper container in the storage closet next to the water heater. The water heater was manufactured by Best Appliance in 1992, model ABC2C833T-LPG, Serial #9219305481. Proper storage of flammable liquids was discussed with the owner, Jason Smith. Loss of contents of the storage closet is valued at $450.00. The engine company returned to quarters at 13:58, with a recovery time of one hour to refill engine water tank, reload hose, change out air bottles, and complete the Fire Incident Report.

2. On October 12, 1995 at 01:55, a three-person engine company from Station #1 was dispatched to a vehicle fire, incident number 3097, in an open field located at Birch Road and Oak Street. There had been several stolen vehicles burned in the area within the last four weeks. It appeared that the vehicles were being stolen in a nearby town and driven to River View, where they were being stripped and burned. Although the call was only 1 mile from the station, the engine company arrived on scene four minutes later due to difficult off-road access to the location. Upon arrival, they found two small pickup trucks parked end to end. The 1994 Chevrolet in the back was fully involved with fire. The preconnected 150 foot 1½ inch line was pulled and the fire was extinguished with 500 gallons of water. Fire was controlled at 2:07. While the fire was being extinguished, further investigation by the captain revealed a line of damp soil where gasoline had been poured, from the fire-involved vehicle to the one in front of it. The interior of the second vehicle was drenched in gasoline, with no fire. The license plate of the burned vehicle was CYN 032. The engine company returned to quarters at 03:15 and had 1½ hours of recovery time to put the engine back into service and complete paperwork.

FIRE DEPARTMENT INCIDENT REPORT

GENERAL INFORMATION
YEAR/INCIDENT NUMBER 9 5 - 3 0 6 2   INCIDENT DATE 1 0 0 3 9 5
DISPATCH TIME 1 5 3 8   END TIME 1 5 5 1   ALARM SOURCE 7
SITUATION FOUND 1 5   PROPERTY MANAGEMENT 3
INCIDENT ADDRESS/LOCATION Park Place and Mountain Blvd
CITY River View   GENERAL PROPERTY USE 1 1
SPECIFIC PROPERTY USE 9 3 8   OCCUPANCY TYPE _ _ . _
STRUCTURED OCCUPIED AT TIME OF INCIDENT _
OWNER NAME River View City
OWNER ADDRESS 1100 Main Street, River View
OCCUPANT NAME

FOR MOBILE PROPERTY INVOLVED
TYPE _ _   LICENSE NUMBER _ _ _ _ _ _ _   YEAR _ _
MAKE _ _ _ _ _ _ _ _ _ _ _ _ _ _   MODEL _ _ _ _ _ _ _ _ _ _ _ _ _ _

COMPLETE FOR ALL FIRES
ACTION TAKEN 1 5   FIRE ORIGIN AREA 9 4   LEVEL A 0 1   FORM OF HEAT 6 1
IGNITION FACTOR 2 1   METHOD OF EXTINGUISHMENT 5
MATERIAL IGNITED FORM 7 4   TYPE 0 1   CONTRIBUTING FACTORS 2 1 2
PROPERTY LOSS _ _ _ _ _ _ _ _ 0   CONTENTS LOSS _ _ _ _ _ _ _ _ _ _
ACRES BURNED _ _ _ 0 . 1   FIRE CONTROLLED DATE _ _ _ _ _ _ _   TIME 1 5 4 2

IF EQUIPMENT INVOLVED
TYPE _ _   MODEL _
SERIAL NUMBER _

COMPLETE FOR ALL STRUCTURE FIRES
CONSTRUCTION TYPE _   ROOF COVERING _   NUMBER OF STORIES _ _
EXTENT OF DAMAGE FLAME _   SMOKE _   SMOKE GENERATION TYPE _ _   FORM _ _

APPARATUS AND PERSONNEL

UNIT RESPONSE	NUMBER PEOPLE	MILES ONE WAY	DISPATCH DATE/TIME	ARRIVAL DATE/TIME	RETURN DATE/TIME	RECOV TIME
E 1	0 0 3	0 0 1	1 0 0 3 9 5 1 5 3 8	1 0 0 3 9 5 1 5 4 0	1 0 0 3 9 5 1 5 5 1	3 0

COMMENTS: Grass fire in park, small spot. Appeared to have been caused by juveniles smoking.

ACTIONS TAKEN 1 5
SIGNATURE: _____   DATE: _____

Figure 10.2

Assignments

FIRE PREVENTION ACTIVITIES

Of the 5,700 people who die annually in fires, nearly 80% of those deaths occur in the home. Volunteering time for local fire prevention programs and activities are ways for the person interested in becoming a firefighter to gain experience and begin saving lives and property through education. Many federal, state, and local fire departments use volunteers to support their fire prevention programs. Volunteers in these programs are involved in a wide range of prevention programs and activities. Some of these include school programs, fair booths, health fairs, group presentations, and initial inspection of local properties for weed abatement problems. In addition to using volunteers for fire prevention activities, many agencies are combining resources and forming interagency fire prevention cooperatives. A fire prevention booth at the county fair may now be staffed by employees of more than one fire agency. For the volunteer, this offers exposure to different types of fire prevention materials relating to specific hazards encountered in other agencies' response areas, experience working in an interagency climate, and networking with other people in the profession.

1. Contact the local fire agencies for information on the types of fire prevention programs and activities they are involved in. Some of the agencies, which may be applicable to your area, could include city, county, state divisions of forestry, and federal fire agencies. Federal agencies include the U.S. Forest Service, under the Department of Agriculture, and Bureau of Land Management, under the Department of the Interior. Both of these agencies are divided into geographical regions and districts. While each district may have some differences, volunteers are used by both of these departments. Not only does the volunteer gain experience with the Forest Service, but the time volunteered is considered specialized training, which is creditable when applying for seasonal jobs with them. Some state, county, and city programs also give credit on job applications for volunteer work. Ask each agency you contact if they have a program that utilizes volunteers from the community for support.

   _____

   _____

   _____

2. If there is more than one fire agency in your geographic area, find out if there are any interagency fire prevention efforts and what they consist of.

   _____

   _____

   _____

3. Participate in a fire prevention program. Even if there is no formal volunteer program in place in the local department, you may be able to accompany an engine company to a school program or help with a station tour.

EXCERPTS FROM NFPA 901 UNIFORM CODING

Alarm Source

1. Telephone direct to fire department
2. Municipal fire alarm system
4. Radio
7. Telephone tie in to fire department, including 911 calls

Situation Found

11. Structure fire
12. Fire in mobil property when used as a structure
14. Vehicle fire
15. Fire in trees, brush, grass, standing crops

Property Management

1. Private tax-paying property
3. City, town, village or other local government property
5. State government property, except military
6. Federal government property, except military

General Use

11. Public recreation use
41. One or two family residential use
65. Farm, agricultural use
93. Wild land

Structure Type

1. Building with one specific property use
2. Building with two or more specific property uses
3. Open structure
4. Open platform

Structure Status

1. Under construction
2. In use with furnishings in place and the property being routinely used
3. Vacant but property secured and maintained
4. Abandoned with property unsecured and not maintained

Specific Property Use

411 One-family dwelling; year-round use

655 Crops, orchard

815 Barns, stables

938 Graded and cared for plots of land, included are parks

Occupancy Type

A1. An assembly building or portion of a building having a stage and an occupant load of 1,000 or more

B2. Drinking or dining establishments having an occupant load of less than 50, wholesale and retail stores, office buildings, printing plants, municipal police and fire stations, factories and workshops using material not highly flammable or combustible, storage and sales rooms for combustible goods, paint stores without bulk handling

E1. Any building used for educational purposes through the 12th grade by less than 50 persons for more than 12 hours per week or four hours in any one day

R3. Dwellings and lodging houses

Structure Occupied at Time of Incident

1. Structure or vehicle occupied at time of incident
2. Structure or vehicle not occupied at time of incident
3. Unit unoccupied, but structure occupied

Actions Taken

11. Rescue, ventilation, extinguishment, salvage, and overhaul
13. Extinguishment, salvage, and overhaul
15. Extinguishment
17. Establish wildfire fire lines

Fire Origin: Area

01. Hallway, corridor, mall

43. Supply storage room or area

81. Passenger area of transportation equipment

94. Lawn, field, open area, included are parks and vacant lots

Level

A. Above ground

B. Below ground

1. 10′ or one building story

2. 20′ or two building stories

A01 = Grade or 1st floor

Form of Heat

01. Outside open fire for debris or waste disposal

15. Heat from natural gas fueled equipment other than torch

61. Cigarette

64. Match

Ignition Factor

11. Incendiary, Arson, Criminal Act

21. Reckless, failure to use ordinary care.

41. Flammable liquid or gas spilled, released accidentally

71. Collision, overturn, knockdown

Method of Extinguishment

1. Self-extinguished

3. Portable extinguisher

4. Automatic extinguishing system

5. Water carried on apparatus initially assigned to the incident

Material Ignited: Form

11. Exterior roof covering, surface, finish

65. Fuel, included are flammable liquids or gases in their final container prior to direct transfer into the engine or burner or the piping associated with this final transfer

74. Growing or natural form whether living or dead, included are forests, brush, and grass

86. Gas or liquid in or from pipe or container

Type

01. Grass

11. Natural gas

23. Gasoline

31. Fat, grease (food)

Contributing Factors

112. Roof collapse

212. Careless act

263. Crime cover: burglary, theft, other

316. Storage: improper

Equipment Involved: Type

01. Road transport vehicle

07. Railroad vehicle

11. Central heating unit

12. Water heater

Construction Type

1. Type I, previously called Fire Resistive

2. Type II, previously called Noncombustible

3. Type III, previously called Ordinary

4. Type IV, previously called Heavy Timber

Roof Covering

1. Tile (clay, cement, slate, etc.)
2. Composition shingles
3. Wood shakes or shingles
5. Metal

Extent of Damage: Flame & Smoke

1. Confined to the object of origin
2. Confined to part of room or area of origin
3. Confined to room of origin
4. Extended beyond structure of origin

Smoke Generation: Type

01. Grass
11. Natural gas
23. Gasoline
31. Fat, grease (food)

Form

11. Exterior roof covering, surface, finish
64. Tire
75. Rubbish, trash, waste
86. Gas or liquid in or from pipe or container

**FIRE DEPARTMENT
INSPECTION REPORT**

DATE: _____

BUSINESS NAME: _____

GENERAL USE: _____

OCCUPANCY CLASS: ___ ___   PROPERTY CLASS: ___ ___ ___

LOCATION: _____
               Street Address   City

NUMBER OF BLDGS: ___   NO. OF STORIES: ___   BASEMENT (B) ___
                                                        CELLAR   (C) ___

MANAGER: _____   PHONE: _____

OWNER: _____   PHONE: _____

DISPOSITION

1-CORRECTED   2-WILL CORRECT   3-VIOLATION NOTICE ISSUED   4-CITATION
                                                            CALL BACK NECESSARY   ISSUED

BUILDING	YES/NA	NO/DISP	FIRE PROTECTION EQUIP	YES/NA	NO/DISP
1. EXIT DOORS	/	/	8. FIRE EXTINGUISHERS	/	/
2. EXIT SIGNS	/	/	9. AUTOMATIC SYSTEMS	/	/
3. EXIT CORRIDORS	/	/	10. WET STANDPIPES	/	/
4. AISLE SPACING	/	/	11. DRY STANDPIPES	/	/
5. OCCUPANT LOAD SIGN	/	/	12. ALARM SYSTEMS	/	/
6. VERTICAL OPENINGS	/	/	13. FIRE ASSEMBLIES	/	/
7. WARNING SIGNS	/	/	14. FIRE WALLS	/	/
COMMON HAZARDS	YES/NA	NO/DISP	**SPECIAL HAZARDS**	YES/NA	NO/DISP
15. ELECTRICAL	/	/	22. GREASE HOODS–DUCTS	/	/
16. FURNACE/BOILER RM	/	/	23. FLAMMABLE LIQUIDS	/	/
17. COOKING EQUIPMENT	/	/	24. LIQUID PROPANE GAS	/	/
18. HEATING EQUIPMENT	/	/	25. COMPRESSED GAS	/	/
19. DECORATIONS	/	/	26. CHEMICALS	/	/
20. HOUSEKEEPING	/	/			
21. WEEDS	/	/			

REMARKS: _____
_____
_____
_____

CLEARANCE: GRANTED (G) ____   INSPECTOR _____

                DENIED (D) ____   OWNER/MANAGER _____

FIRE DEPARTMENT
INSPECTION REPORT

DATE: _____

BUSINESS NAME: _____

GENERAL USE: _____

OCCUPANCY CLASS: ____ ____   PROPERTY CLASS: ____ ____ ____

LOCATION: _____
             Street Address   City

NUMBER OF BLDGS: ____   NO. OF STORIES: ____   BASEMENT (B) ____
                                               CELLAR   (C)

MANAGER: _____   PHONE: _____

OWNER: _____   PHONE: _____

DISPOSITION

| 1-CORRECTED | 2-WILL CORRECT | 3-VIOLATION NOTICE ISSUED
CALL BACK NECESSARY | 4-CITATION
ISSUED |

BUILDING	YES/NA	NO/DISP	FIRE PROTECTION EQUIP	YES/NA	NO/DISP
1. EXIT DOORS	/	/	8. FIRE EXTINGUISHERS	/	/
2. EXIT SIGNS	/	/	9. AUTOMATIC SYSTEMS	/	/
3. EXIT CORRIDORS	/	/	10. WET STANDPIPES	/	/
4. AISLE SPACING	/	/	11. DRY STANDPIPES	/	/
5. OCCUPANT LOAD SIGN	/	/	12. ALARM SYSTEMS	/	/
6. VERTICAL OPENINGS	/	/	13. FIRE ASSEMBLIES	/	/
7. WARNING SIGNS	/	/	14. FIRE WALLS	/	/
COMMON HAZARDS	YES/NA	NO/DISP	**SPECIAL HAZARDS**	YES/NA	NO/DISP
15. ELECTRICAL	/	/	22. GREASE HOODS–DUCTS	/	/
16. FURNACE/BOILER RM	/	/	23. FLAMMABLE LIQUIDS	/	/
17. COOKING EQUIPMENT	/	/	24. LIQUID PROPANE GAS	/	/
18. HEATING EQUIPMENT	/	/	25. COMPRESSED GAS	/	/
19. DECORATIONS	/	/	26. CHEMICALS	/	/
20. HOUSEKEEPING	/	/			
21. WEEDS	/	/			

REMARKS: _____
_____
_____
_____

CLEARANCE: GRANTED (G) ____   INSPECTOR _____

           DENIED   (D) ____   OWNER/MANAGER _____

**FIRE DEPARTMENT
INSPECTION REPORT**

DATE: _____

BUSINESS NAME: _____

GENERAL USE: _____

OCCUPANCY CLASS: ___ ___   PROPERTY CLASS: ___ ___ ___

LOCATION: _____
Street Address   City

NUMBER OF BLDGS: ____ NO. OF STORIES: ____ BASEMENT (B) ____
                                            CELLAR   (C)

MANAGER: _____ PHONE: _____

OWNER: _____ PHONE: _____

DISPOSITION

| 1-CORRECTED | 2-WILL CORRECT | 3-VIOLATION NOTICE ISSUED CALL BACK NECESSARY | 4-CITATION ISSUED |

BUILDING	YES/NA	NO/DISP	FIRE PROTECTION EQUIP	YES/NA	NO/DISP
1. EXIT DOORS	/	/	8. FIRE EXTINGUISHERS	/	/
2. EXIT SIGNS	/	/	9. AUTOMATIC SYSTEMS	/	/
3. EXIT CORRIDORS	/	/	10. WET STANDPIPES	/	/
4. AISLE SPACING	/	/	11. DRY STANDPIPES	/	/
5. OCCUPANT LOAD SIGN	/	/	12. ALARM SYSTEMS	/	/
6. VERTICAL OPENINGS	/	/	13. FIRE ASSEMBLIES	/	/
7. WARNING SIGNS	/	/	14. FIRE WALLS	/	/
COMMON HAZARDS	YES/NA	NO/DISP	**SPECIAL HAZARDS**	YES/NA	NO/DISP
15. ELECTRICAL	/	/	22. GREASE HOODS–DUCTS	/	/
16. FURNACE/BOILER RM	/	/	23. FLAMMABLE LIQUIDS	/	/
17. COOKING EQUIPMENT	/	/	24. LIQUID PROPANE GAS	/	/
18. HEATING EQUIPMENT	/	/	25. COMPRESSED GAS	/	/
19. DECORATIONS	/	/	26. CHEMICALS	/	/
20. HOUSEKEEPING	/	/			
21. WEEDS	/	/			

REMARKS: _____
_____
_____

CLEARANCE: GRANTED (G) ____   INSPECTOR _____

            DENIED (D) ____   OWNER/MANAGER _____

FIRE DEPARTMENT
INSPECTION REPORT

DATE: _____

BUSINESS NAME: _____

GENERAL USE: _____

OCCUPANCY CLASS: ___ ___   PROPERTY CLASS: ___ ___ ___

LOCATION: _____
              Street Address   City

NUMBER OF BLDGS: ____ NO. OF STORIES: ____ BASEMENT (B) ____
                                                        CELLAR (C) ____

MANAGER: _____ PHONE: _____

OWNER: _____ PHONE: _____

DISPOSITION

1-CORRECTED   2-WILL CORRECT   3-VIOLATION NOTICE ISSUED   4-CITATION
                                                                     CALL BACK NECESSARY   ISSUED

BUILDING	YES/NA	NO/DISP
1. EXIT DOORS	/	/
2. EXIT SIGNS	/	/
3. EXIT CORRIDORS	/	/
4. AISLE SPACING	/	/
5. OCCUPANT LOAD SIGN	/	/
6. VERTICAL OPENINGS	/	/
7. WARNING SIGNS	/	/

COMMON HAZARDS	YES/NA	NO/DISP
15. ELECTRICAL	/	/
16. FURNACE/BOILER RM	/	/
17. COOKING EQUIPMENT	/	/
18. HEATING EQUIPMENT	/	/
19. DECORATIONS	/	/
20. HOUSEKEEPING	/	/
21. WEEDS	/	/

FIRE PROTECTION EQUIP	YES/NA	NO/DISP
8. FIRE EXTINGUISHERS	/	/
9. AUTOMATIC SYSTEMS	/	/
10. WET STANDPIPES	/	/
11. DRY STANDPIPES	/	/
12. ALARM SYSTEMS	/	/
13. FIRE ASSEMBLIES	/	/
14. FIRE WALLS	/	/

SPECIAL HAZARDS	YES/NA	NO/DISP
22. GREASE HOODS–DUCTS	/	/
23. FLAMMABLE LIQUIDS	/	/
24. LIQUID PROPANE GAS	/	/
25. COMPRESSED GAS	/	/
26. CHEMICALS	/	/

REMARKS: _____
_____
_____
_____

CLEARANCE: GRANTED (G) ____   INSPECTOR _____

                 DENIED   (D) ____   OWNER/MANAGER _____

FIRE DEPARTMENT
INSPECTION REPORT

DATE: _____

BUSINESS NAME: _____

GENERAL USE: _____

OCCUPANCY CLASS: ___ ___   PROPERTY CLASS: ___ ___ ___

LOCATION: _____
                 Street Address   City

NUMBER OF BLDGS: ___ NO. OF STORIES: ___ BASEMENT (B) ___
                                         CELLAR   (C) ___

MANAGER: _____ PHONE: _____

OWNER: _____ PHONE: _____

DISPOSITION

1-CORRECTED   2-WILL CORRECT   3-VIOLATION NOTICE ISSUED   4-CITATION
                                                       CALL BACK NECESSARY   ISSUED

BUILDING	YES/NA	NO/DISP	FIRE PROTECTION EQUIP	YES/NA	NO/DISP
1. EXIT DOORS	/	/	8. FIRE EXTINGUISHERS	/	/
2. EXIT SIGNS	/	/	9. AUTOMATIC SYSTEMS	/	/
3. EXIT CORRIDORS	/	/	10. WET STANDPIPES	/	/
4. AISLE SPACING	/	/	11. DRY STANDPIPES	/	/
5. OCCUPANT LOAD SIGN	/	/	12. ALARM SYSTEMS	/	/
6. VERTICAL OPENINGS	/	/	13. FIRE ASSEMBLIES	/	/
7. WARNING SIGNS	/	/	14. FIRE WALLS	/	/
COMMON HAZARDS	YES/NA	NO/DISP	**SPECIAL HAZARDS**	YES/NA	NO/DISP
15. ELECTRICAL	/	/	22. GREASE HOODS–DUCTS	/	/
16. FURNACE/BOILER RM	/	/	23. FLAMMABLE LIQUIDS	/	/
17. COOKING EQUIPMENT	/	/	24. LIQUID PROPANE GAS	/	/
18. HEATING EQUIPMENT	/	/	25. COMPRESSED GAS	/	/
19. DECORATIONS	/	/	26. CHEMICALS	/	/
20. HOUSEKEEPING	/	/			
21. WEEDS	/	/			

REMARKS: _____
_____
_____
_____

CLEARANCE: GRANTED (G) ____   INSPECTOR _____

               DENIED (D) ____   OWNER/MANAGER _____

FIRE DEPARTMENT
INSPECTION REPORT

DATE: _____

BUSINESS NAME: _____

GENERAL USE: _____

OCCUPANCY CLASS: ___ ___   PROPERTY CLASS: ___ ___ ___

LOCATION: _____
               Street Address   City

NUMBER OF BLDGS: ____ NO. OF STORIES: ____ BASEMENT (B) ____
                                                    CELLAR   (C)

MANAGER: _____ PHONE: _____

OWNER: _____ PHONE: _____

DISPOSITION

1-CORRECTED   2-WILL CORRECT   3-VIOLATION NOTICE ISSUED   4-CITATION
                                                      CALL BACK NECESSARY   ISSUED

BUILDING	YES/NA	NO/DISP	FIRE PROTECTION EQUIP	YES/NA	NO/DISP
1. EXIT DOORS	/	/	8. FIRE EXTINGUISHERS	/	/
2. EXIT SIGNS	/	/	9. AUTOMATIC SYSTEMS	/	/
3. EXIT CORRIDORS	/	/	10. WET STANDPIPES	/	/
4. AISLE SPACING	/	/	11. DRY STANDPIPES	/	/
5. OCCUPANT LOAD SIGN	/	/	12. ALARM SYSTEMS	/	/
6. VERTICAL OPENINGS	/	/	13. FIRE ASSEMBLIES	/	/
7. WARNING SIGNS	/	/	14. FIRE WALLS	/	/
COMMON HAZARDS	YES/NA	NO/DISP	SPECIAL HAZARDS	YES/NA	NO/DISP
15. ELECTRICAL	/	/	22. GREASE HOODS–DUCTS	/	/
16. FURNACE/BOILER RM	/	/	23. FLAMMABLE LIQUIDS	/	/
17. COOKING EQUIPMENT	/	/	24. LIQUID PROPANE GAS	/	/
18. HEATING EQUIPMENT	/	/	25. COMPRESSED GAS	/	/
19. DECORATIONS	/	/	26. CHEMICALS	/	/
20. HOUSEKEEPING	/	/			
21. WEEDS	/	/			

REMARKS: _____
_____
_____
_____

CLEARANCE: GRANTED (G) ____   INSPECTOR _____

               DENIED (D) ____   OWNER/MANAGER _____

FIRE DEPARTMENT INCIDENT REPORT

GENERAL INFORMATION
YEAR/INCIDENT NUMBER __ __ - __ __ __ __ __ __   INCIDENT DATE __ __ __ __ __ __

DISPATCH TIME __ __ __ __   END TIME __ __ __ __   ALARM SOURCE __

SITUATION FOUND __ __   PROPERTY MANAGEMENT __

INCIDENT ADDRESS/LOCATION _____

CITY _____   GENERAL PROPERTY USE __ __

SPECIFIC PROPERTY USE __ __ __   OCCUPANCY TYPE __ __ . __

STRUCTURED OCCUPIED AT TIME OF INCIDENT __

OWNER NAME _____

OWNER ADDRESS _____

OCCUPANT NAME _____

FOR MOBILE PROPERTY INVOLVED
TYPE __ __   LICENSE NUMBER __ __ __ __ __ __ __ __   YEAR __ __

MAKE __ __ __ __ __ __ __ __ __ __ __ __ __ __ __   MODEL __ __ __ __ __ __ __ __ __ __ __ __ __ __ __

COMPLETE FOR ALL FIRES
ACTION TAKEN __ __   FIRE ORIGIN AREA __ __   LEVEL __ __ __   FORM OF HEAT __ __

IGNITION FACTOR __ __   METHOD OF EXTINGUISHMENT __

MATERIAL IGNITED FORM __ __   TYPE __ __   CONTRIBUTING FACTORS __ __ __

PROPERTY LOSS __ __ __ __ __ __ __ __ __   CONTENTS LOSS __ __ __ __ __ __ __ __ __

ACRES BURNED __ __ __ __ . __   FIRE CONTROLLED DATE __ __ __ __ __ __   TIME __ __ __ __

IF EQUIPMENT INVOLVED
TYPE __ __   MODEL __

SERIAL NUMBER __

COMPLETE FOR ALL STRUCTURE FIRES
CONSTRUCTION TYPE __   ROOF COVERING __   NUMBER OF STORIES __ __

EXTENT OF DAMAGE FLAME __   SMOKE __   SMOKE GENERATION TYPE __ __   FORM __ __

APPARATUS AND PERSONNEL

UNIT RESPONSE	NUMBER PEOPLE	MILES ONE WAY	DISPATCH DATE/TIME	ARRIVAL DATE/TIME	RETURN DATE/TIME	RECOV TIME
____	____	____	_____ ____	_____ ____	_____ ____	___

COMMENTS: _____

_____

ACTIONS TAKEN __ __

SIGNATURE: _____   DATE: _____

FIRE DEPARTMENT INCIDENT REPORT

GENERAL INFORMATION
YEAR/INCIDENT NUMBER __ __ - __ __ __ __ __ __   INCIDENT DATE __ __ __ __ __ __

DISPATCH TIME __ __ __ __   END TIME __ __ __ __   ALARM SOURCE __

SITUATION FOUND __ __   PROPERTY MANAGEMENT __

INCIDENT ADDRESS/LOCATION _____

CITY _____   GENERAL PROPERTY USE __ __

SPECIFIC PROPERTY USE __ __ __   OCCUPANCY TYPE __ __ . __

STRUCTURED OCCUPIED AT TIME OF INCIDENT __

OWNER NAME _____

OWNER ADDRESS _____

OCCUPANT NAME _____

FOR MOBILE PROPERTY INVOLVED
TYPE __ __   LICENSE NUMBER __ __ __ __ __ __ __   YEAR __ __

MAKE __ __ __ __ __ __ __ __ __ __ __ __ __   MODEL __ __ __ __ __ __ __ __ __ __ __ __ __

COMPLETE FOR ALL FIRES
ACTION TAKEN __ __   FIRE ORIGIN AREA __ __   LEVEL __ __ __   FORM OF HEAT __ __

IGNITION FACTOR __ __   METHOD OF EXTINGUISHMENT __

MATERIAL IGNITED FORM __ __   TYPE __ __   CONTRIBUTING FACTORS __ __ __

PROPERTY LOSS __ __ __ __ __ __ __ __ __   CONTENTS LOSS __ __ __ __ __ __ __ __ __

ACRES BURNED __ __ __ __ __   FIRE CONTROLLED DATE __ __ __ __ __ __   TIME __ __ __ __

IF EQUIPMENT INVOLVED
TYPE __ __   MODEL __

SERIAL NUMBER __

COMPLETE FOR ALL STRUCTURE FIRES
CONSTRUCTION TYPE __   ROOF COVERING __   NUMBER OF STORIES __ __

EXTENT OF DAMAGE FLAME __   SMOKE __   SMOKE GENERATION TYPE __ __   FORM __ __

APPARATUS AND PERSONNEL

UNIT RESPONSE	NUMBER PEOPLE	MILES ONE WAY	DISPATCH DATE/TIME	ARRIVAL DATE/TIME	RETURN DATE/TIME	RECOV TIME
_ _ _ _	_ _ _	_ _ _	_ _ _ _ _ _	_ _ _ _ _ _	_ _ _ _ _ _	_ _ _

COMMENTS: _____

_____

ACTIONS TAKEN __ __

SIGNATURE: _____   DATE: _____

Chapter 11

Codes and Ordinances

"Reason is the life of the law."
Sir Edward Coke, 1552–1634

INTRODUCTION

Model fire codes are a collection of laws that are organized in such a way that they can be easily understood and used. The greatest motivation for the development and modification of fire codes has been from fires with high life and property losses. This has been true from the fire loss in Colonial America that precipitated the 1648 Nieuw Amsterdam law prohibiting construction of wood or plaster chimneys; to the California fire storm losses in 1993 that led to the law prohibiting construction of roofs with wood shake shingles.

The Model Fire Codes in use today are developed by committees of experts who represent the local communities in which the codes are adopted, and the national fire community.

Questions

1. In regard to fire code violations, the fire department employee conducting the fire safety inspection has the final word on determining the intent of the law.
   True   False

2. Local ordinances come under the definition of statutory laws.
   True   False

3. Model national codes are the basis for most state fire prevention laws.
   True   False

4. Model codes when adopted by the jurisdiction have the force of law.
   True   False

5. As a public safety employee, you cannot be sued civilly for actions taken in the line of duty.
   True   False

6. According to most state vehicle codes, any vehicle driven by a qualified firefighter to an emergency is considered to be an authorized emergency vehicle.
   True   False

7. The intent of a law is interpreted in court.
   True   False

8. Codes are often developed in response to disasters and the recognition of the need for public safety.
   True   False

9. Standards may be adopted by ordinance or department policy.
   True   False

10. As public safety officers, firefighters are allowed to issue citations in areas outside of their own jurisdiction.
    True   False

11. The Standard, Uniform, and Basic Model Fire Prevention Codes are companion codes to Building Codes available through the same publisher.
    True   False

12. Some violations encountered during fire department operations and inspections will not be fire code violations; those violations will be referred to other departments such as the building and health departments.
    True   False

13. Which of the following is **not** a way to avoid lawsuits?
    A. Pay attention to your training
    B. Practice what you learn
    C. Document training
    D. Perform to the level to which you have been trained
    E. All of the above are ways to avoid lawsuits

14. Which of the following is **not** true in regard to fire safety inspections of commercial premises?
    A. Inspectors should be adequately identified
    B. The reason for inspection must be stated by the inspector
    C. The inspector may inspect the complete premises
    D. The inspector may issue stop orders for extremely hazardous conditions

15. When the agency designated by law to take charge of emergency scenes is not present, who assumes control of the scene?
    A. Highest ranking member of the public agency at scene
    B. State police
    C. Fire department
    D. Local law enforcement

16. Model codes are adopted by most jurisdictions through
    A. the state fire marshal.
    B. local ordinance.
    C. state mandate.
    D. local fire department.

17. Unserviced fire extinguishers at a business would be an example of a
    A. physical illegality.
    B. procedural illegality.
    C. structural deficiency.

18. According to Federal OSHA standards, what is the minimum number of individuals, in full personal protective equipment, required for entrance into an atmosphere immediately hazardous to life and health?
    A. 2
    B. 3
    C. 4
    D. 5

Match the following actions, which can result in a tort used in a civil suit, to the definition that best describes them.

_____ 19. Nonfeasance
_____ 20. Misfeasance
_____ 21. Malfeasance

A. Breaking a law or misconduct.
B. Doing something wrong within the framework of the law.
C. Failure to act.

Match the Model Fire Code to the agency that publishes it.

_____ 22. Standard Fire Fire Prevention Code

_____ 23. Uniform Fire Code

_____ 24. National Fire Codes

_____ 25. Basic Fire Prevention Code

A. National Fire Prevention Code

B. Building Officials & Code Administrators

C. Southern Building Code Congress International

D. International Fire Code Institute

Exercises

Exercise 1

To comply with federal safety regulations, what action could be taken by a three-person engine company at a free burning structure fire?

_____

_____

Exercise 2

Read the account of the incident and answer the questions that follow.

A fire on December 30, 1903 at the Iroquois Theater in Chicago claimed the lives of 602 people. As a result, many fire regulations were passed in an attempt to prevent another tragedy of this type. The fire originated when a light blew a fuse and ignited the highly flammable painted backdrops. An asbestos curtain designed to separate the burning stage from the seating area of the theater became stuck and did not lower completely. The number of people in the theater exceeded the exiting capacity of the aisle ways. In the ensuing panic, many people were trampled to death. Those who did make it to the exits found doors that opened inward blocked by the mass of people trying to get out. Bodies were found stacked several deep in front of those doors.

List three areas where new regulations governing different aspects of the theater could help avoid another tragedy of this type.

1. _____

2. _____

3. _____

Look up the three areas you chose in the Model Fire Code of your choice and list the assembly requirements for those areas.

1. _____
   _____

2. _____
   _____

3. _____
   _____

Assignments

1. Which fire code has been adopted by the local fire department or the fire department in the jurisdiction of your choice?

   _____

   _____

2. What are the fire extinguisher requirements for a business of less than 75 square feet in the model code you listed above.

   Number of extinguishers required: _____

   Required placement of extinguishers: _____

3. Have the National Fire Protection Association's (NFPA) Model Fire Codes been adopted by the jurisdiction of your choice? If yes, was it adopted by ordinance or by policy?

   _____

4. Is there a formal complaint procedure in the department of your choice? If so, briefly list the steps involved.

   _____

   _____

   _____

   _____

5. According to your State Vehicle Code, what is the minimum equipment requirement for a vehicle to be considered an emergency vehicle?

   _____

   _____

6. According to your State Vehicle Code, is there a code specifying the following distance required between emergency vehicles when responding code 3 to an emergency? If so, what is it?

   _____

Fire Protection Systems and Equipment

"Built-in fire protection, which is on duty 24 hours a day, 365 days a year, and is never busy on another assignment."

Robert Klinoff, Fire Marshal
Kern County Fire Department

INTRODUCTION

Advances in fire protection systems and equipment have contributed to an increase in life safety and a decrease in property loss. Among these advances are better designed water systems, standpipe systems, and sprinkler systems. Sprinkler systems discharge water directly on the fire while it is small. They are generally 96 percent effective in extinguishing or controlling a fire. When the fire is not extinguished by the sprinklers, the rate of spread is decreased and the products of combustion are limited until fire crews arrive. Thirty-five percent of fires in sprinklered buildings are extinguished or controlled by one sprinkler head, with the majority of fires extinguished or controlled by five or less sprinkler heads. Residential sprinklers are now becoming more popular. Combined with smoke alarms, they should help to reduce the 80 percent residential fatality rate in our country.

Standpipe systems reduce the time it takes firefighters to put hose streams into operation on multi-story structures. The first priority for an engine operator is to establish an adequate water supply for the operation. In most areas, buildings over four stories high require a standpipe system.

Preincident planning is very important in determining the types of fire protection systems and equipment in use in your response area. Being familiar, before the fire, with the location of fire department connections on standpipe systems, the location of fire hydrants, and the procedure for supplying the system greatly increases fire operation effectiveness.

Questions

1. A pitot gauge is used to test the fire flow capability of a fire hydrant.
   True   False

2. Yearly service and maintenance of fire hydrants is usually performed by the local water company.
   True   False

3. In water system mains, it is recommended that valves be placed a maximum of 800 feet apart to prevent collapse of the water system during times of excessive use.
   True   False

4. Because of the large size of water system supply lines, there is no danger of water hammer when closing a fire hydrant valve.
   True   False

5. Long-term retardants are applied from the air during direct attack on wildland fires.
   True   False

6. Automatic sprinklers control 96% of the fires where they are activated.
   True   False

7. A wet pipe sprinkler system would be used in buildings where there is no danger of freezing.
   True   False

8. It is necessary to have written working agreements with the water company that supplies the fire hydrants in your area.
   True   False

9. A gravity fed water system is one in which the
   A. water source is stored at a higher elevation than the city it supplies.
   B. water source is a river.
   C. water source is a great distance from the service area.
   D. water source is a well.

10. What is the most common extinguishing agent for fire?
    A. Water
    B. Foam
    C. Carbon dioxide
    D. Dry chemical

11. What size water main is recommended for water supply at shopping centers and industrial areas?
    A. 4 inches and 6 inches
    B. 6 inches and 8 inches
    C. 8 inches and 12 inches
    D. 12 inches and 16 inches

12. Water hammer is the term used to describe the shock wave created by water when
    A. a straight stream is used to break out a window in a structure.
    B. atmospheric pressure is used on a static water source to force water up into a suction hose.
    C. centrifugal pumps on engines are used in series to increase pumping pressures.
    D. a nozzle or valve is closed too quickly.

13. In which phase of fire are private fire protection systems designed to alert the occupants or fire department?
    A. Incipient
    B. Free burning
    C. Smoldering

14. What is the expansion rate of water to steam?
    A. 1 to 100
    B. 1 to 500
    C. 1 to 1200
    D. 1 to 1700

15. In a dry pipe sprinkler system, what usually keeps the clapper in the main valve body from opening and letting in the water?
    A. Water
    B. Vacuum
    C. Carbon dioxide
    D. Compressed air

16. A system where the sprinkler heads are open all of the time is called
    A. a deluge system.
    B. an open system.
    C. a pre-action system.
    D. a dry system.

17. When a sprinkler system does not activate properly, it is usually due to
    A. low water pressure.
    B. criminal intent.
    C. closed valves.
    D. human error.

18. Which standpipe system has both the 2½ inch fire department connection and the 1½ inch occupant connection with hose?
    A. Class I
    B. Class II
    C. Class III

19. Alcohol foams were developed for use on what types of fuels?
    A. Hydrocarbons
    B. Polar solvents
    C. Vapor clouds
    D. Vegetation

20. What are the two most common percentages of foam concentrations used?
    A. 2% and 4%
    B. 3% and 6%
    C. 4% and 8%
    D. 5% and 10%

Match the fire suppressants to the description that best describes its characteristics.

_____ 21. Water
_____ 22. Foam
_____ 23. Carbon dioxide
_____ 24. Wetting agents
_____ 25. Halogenated agents
_____ 26. Dry chemicals

A. Reduces surface tension
B. Breaks chemical chain reaction
C. Insulates, cools, forms barrier
D. Cools
E. Smothers

Using the NFPA color coding system, match the fire hydrant cap color to the hydrant flow capability.

_____ 27. Black
_____ 28. Green
_____ 29. Orange
_____ 30. Red
_____ 31. White

A. 499 or less gpm
B. 500 to 999 gpm
C. 1,000 or more gpm
D. Dead end main
E. Out of service

32. When using a foam eductor, what four manufacturer's specifications must be followed to produce good foam?

    1. _____
    2. _____
    3. _____
    4. _____

33. What is the life safety hazard of carbon dioxide extinguishing systems in confined areas?

    _____
    _____

34. In the twinned fire suppression systems used on aircraft fires, how do the foam and dry chemical work together to extinguish the fire?

   Water: _____

   Foam: _____

   CO_2: _____

35. What do the letters AFFF in the synthetic foam A Triple F stand for?

   _____ _____ _____ _____ _____

Exercises

Exercise 1

Identify the two types of sprinkler heads in Figure 12.1.

   1. _____

   2. _____

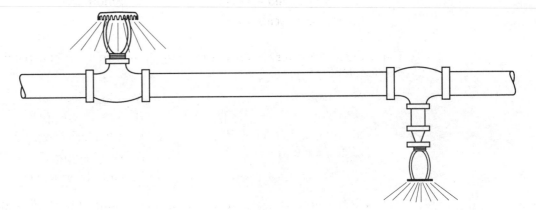

Figure 12.1

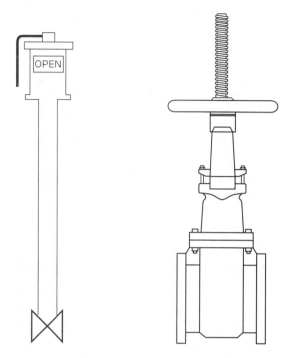

Figure 12.2

Exercise 2

Identify the two types of valves used on sprinkler systems in Figure 12.2.

1. _____

2. _____

What is the purpose of these two types of valves?

_____

Assignments

1. Is the water system that supplies local fire fighting efforts in your area or in the jurisdiction of your choice a public or private water company?

   _____

2. How is the water in the system you selected stored?

   _____

   _____

3. Is the water system a gravity fed, pump, or combination system?

   _____

4. Does the fire department have a written working agreement with the water company? If so, how often is it renewed and what is the department's responsibility in regard to hydrant maintenance?

_____

_____

_____

5. Select a business in your area and briefly describe its fire protection system. Include what class of standpipe system it has and how the sprinkler system is activated.

_____

_____

_____

Chapter 13

Emergency Incident Management

INTRODUCTION

Incident management plays an important part in mitigating an incident in the safest possible manner with the least amount of damage. Emergency incident management has evolved along with the rest of the fire service. During this process, while retaining the unity and chain of command, incident management has become less rank-oriented and is now managed more by function than rank. The five major functional areas of the Incident Command System provide for the maximum utilization of personnel and resources. By managing incidents in this way, the three levels of an incident—strategical, tactical, and task—can be effectively completed.

Management of an incident by function effectively breaks an incident up into manageable parts. By having an operations section with coordinated branches and divisions, span of control becomes more manageable and personnel and equipment can be accounted for. Having personnel assigned to positions they are trained in and knowing they will perform as requested prevents the confusion and lack of accountability that comes from freelancing at an incident.

The foundation for good emergency incident management starts well before the incident. It encompasses our training, experience, prefire plans and inspections, as well as our knowledge of the equipment and water system capabilities in the response area we are assigned to. When the initial call comes in, firefighters begin assessing facts and probabilities based on the dispatch and their knowledge of the area. In fire fighting, the unanticipated probabilities are also assured to occur. Some of these could include a dispatch for a vehicle accident, which upon arrival, turns out to be a vehicle versus a power pole with downed charged electrical lines. Another could be a dispatch for a car fire or medical aid for a person burned, and upon arrival you find the involved vehicle is in a garage or a carport underneath an apartment building.

One thing about emergency incidents is that no two are ever exactly alike. A common system of emergency incident management, which can be expanded or decreased as the incident changes, results in effective fireground operations.

Questions

1. Good communication between forces is essential in a combination attack.
   True   False

2. It is usually not necessary to wear a breathing apparatus during overhaul operations.
   True   False

3. Before implementing ICS, it is necessary to have the five functional areas of command—command, operations, plans, logistics, and finance—staffed and functioning.
   True   False

4. It is not necessary to have a written incident action plan on small incidents.
   True   False

5. On incidents that include large areas and many jurisdictions, it is often necessary to have more than one incident command post.
   True   False

6. On incidents involving multiple jurisdictions, an officer from each jurisdiction involved should be on the unified command team.
   True   False

7. A task force and strike team are a set number of resources of the same kind and type.
   True   False

8. Safety considerations dictate that a defensive strategy should be used on all large fires.
   True   False

9. The jobs required to achieve the incident plan tactics are called
   A. tasks.
   B. divisions.
   C. groups.
   D. strike teams.

10. Which of the following is not a strategic operational mode?
    A. Offensive
    B. Defensive
    C. Ventilation
    D. Combination

11. Verbal radio communications used in the ICS should be in
    A. safety text.
    B. clear text.
    C. fire code.
    D. radio code.

12. In the Incident Command System, what is the standard number of people managed by a single person?
    A. Three
    B. Three to seven
    C. Five
    D. Five to nine

13. In an ICS staging area, available equipment is expected to be able to respond within how many minutes of being called for?
    A. 3 minutes
    B. 5 minutes
    C. 10 minutes
    D. 15 minutes

Match the ICS functional area to the description that best describes it.

_____ 14. Command
_____ 15. Operations
_____ 16. Logistics
_____ 17. Plans
_____ 18. Finance

A. Collects, evaluates, disseminates incident tactical information.
B. Responsible for overall management of the incident.
C. Provides service and support functions for the incident.
D. Deals with vendors, personnel time records, and incident cost analysis.
E. Direct management of all tactical activities at the incident.

19. List the 7 strategic priorities.
    1. _____   5. _____
    2. _____   6. _____
    3. _____   7. _____
    4. _____

20. List the 5 steps in a size up.

   1. _____   4. _____
   2. _____   5. _____
   3. _____

Exercises

The following case studies give an indication of how fireground command has evolved during the last few years. While fire fighting continues to be an inherently dangerous profession, there are many factors that relate to emergency incident management and contribute to a safer operation. Read the case studies and answer the questions that follow them.

Exercise 1

On July 1, 1988, in Hackensack, New Jersey, three firefighters were killed instantly when a wood bowstring truss roof suddenly collapsed. Two firefighters were able to take shelter in the building; however, they died of asphyxiation approximately 13 minutes after the collapse when they ran out of air.

Just after 3:00 in the afternoon, a fire was reported at a car dealership. Upon arrival first-in fire units observed heavy smoke coming from the roof. The fire appeared to be in the attic that ran the length of the vehicle service area. Unknown to the firefighters, the attic space, which consisted of 7,800 square feet, was being used for storage of auto parts as well as oil, gas, cleaning fluids, and antifreeze. This resulted in a significant amount of weight and fuel loading in the attic area. Construction of the service area consisted of a wood bowstring truss roof with concrete block walls.

As a truck company assessed the fire from the roof and started ventilation, firefighters deployed a 1½-inch line into the vehicle service area in an attempt to make entry to the attic. Meanwhile, a second alarm for additional manpower was requested. Operations were hampered by vehicles at the dealership (which were eventually removed) that blocked fire department access to part of the structure. Fireground operations were additionally hampered by the single radio channel available to the fire department. Incident traffic, along with medical aid dispatches, fire dispatches, and call backs for additional manpower were all forced to utilize a single radio channel.

At 15:34:25, as fire activity continued to increase, the battalion chief ordered the interior crews out of the building. According to tapes of the radio traffic there was no answer by the firefighters engaged in interior attack acknowledging receipt of the order to evacuate. Just over two minutes later the roof collapsed killing three firefighters and forcing two firefighters to retreat to a tool storage room.

At 15:37, a radio transmission was recorded from the two firefighters saying they had retreated to a storage closet. The firefighters continued to request help, radioing for help 30 times in the 13 minutes before they ran out of air. People with scanners who heard the requests for help contacted fire headquarters to let them know that there were firefighters trapped and they could hear the bells on their breathing apparatus going off during their transmissions; also that it did not appear that personnel on scene were receiving the transmission. At 15:50:18, the trapped firefighters transmitted that they were out of air. The location of the firefighters was not known and could not be determined. Rescue attempts were unsuccessful.

Exercise 2

On Wednesday, September 25, 1996 at 1:30 P.M., a fire was reported in a cold storage shed at a grape warehouse. Engine companies were dispatched and given a separate radio channel for the incident at that time. Upon arrival they found a cold storage building approximately 330 × 130 square feet, with 50% of the building heavily involved with fire. Two additional alarms for personnel and equipment were subsequently struck. The second alarm went out at 13:27 and the third alarm at 14:01. Additional personnel were needed; however, they had to be requested as single resources. This was due to another incident in progress, which had gone to a second alarm, resulting in no additional engine companies available.

Initial operations consisted of determining what was involved, what the exposures were, and providing a water supply. Exposures included electrical and mechanical rooms with large outside ammonia tanks just past the electrical rooms. The initial incident objective was to confine the fire to the building of origin and keep it away from the ammonia tanks. The private water system was inadequate for the amount of fire and the two water storage tanks on the scene were quickly drained. Water tenders were assigned to the incident for water supply. A rehabilitation area for personnel was set up. Since no additional engine companies were available, personnel on scene were rotated from the fire to rehab and then back to the fire. Individual's vital signs were monitored in rehab to assure they were physically able to return to the fire fighting effort.

The incident was divided into branches with divisions within the branch. Information on fire activity from each area was communicated to the incident commander until an operations officer was assigned to the incident. Two safety officers were assigned to the incident and public information officers were also assigned to handle press releases and news interviews. At 14:16, the safety officer advised the incident commander that there were indications of structure instability of the north wall. At that time, fire fighting personnel were instructed not to place themselves between the north wall and the adjacent structure. A safety issue, in addition to fire fighting efforts, were the efforts of plant personnel involved in trying to recover product from the area. All areas were monitored to ensure safety of everyone on scene. Subsequently, there were collapses of sections of the building.

At 14:56, a supply engine at a hydrant reported that the hydrant was down and their water supply was gone. The water company was contacted with a request to boost water pressure for the area. A crew was sent by the water company to support the incident. This consisted of opening a valve from the public water system to support the private water system at the scene and provide the necessary volume of water. At 15:11, as fire activity increased, the safety officer advised crews on the north side of the building that the ammonia was venting and self-contained breathing apparatus were to be used. An evacuation of the east side was advised by a company officer and the safety officer at that time.

As the incident progressed, the adjusted objectives included: keep fire from crossing breeze way to the west, keep fire from cold storage warehouse on the north, stop fire on east to keep out of mechanical rooms, protect large ammonia tanks on east end. It was also determined by the safety officer that there were other hazardous chemicals present, like sulfur dioxide and possible cyanide in the insulation. The potential for fire spread to exposures and ammonia tanks was continually monitored.

The incident was completed on September 25, 1996, at 22:30. At the height of the incident, there were 70 personnel involved in fire fighting activities. As the fire was controlled, personnel were released or rotated off the incident. Incident objectives were ultimately met. Losses were estimated between five and ten million dollars.

1. Briefly explain how fireground command has evolved from being rank-oriented to function-oriented as it relates to the incidents.

2. How does recognizing that a fire is beyond the current resources available to control it affect incident objectives?

3. List and briefly explain two additional areas where emergency incident management has evolved from the first incident to the second.

   1.

   2.

Assignments

1. Does your local community train together for disasters? If so, which groups or agencies participate and who is responsible for organizing them?

2. Do agencies other than the fire department in your area understand and use the Incident Command System?

3. What is the signal, in the department of your choice, for immediate evacuation of a structure or area?

Chapter 14

Emergency Operations

"Union gives strength."
Aseop, 550 B.C., The Bundle of Sticks

INTRODUCTION

Emergency operations are where all of your training and experience are put into action. Emergency operations must be coordinated from the time the call for help comes in to the completion of it. This includes the initial dispatch of predetermined equipment and personnel for the specific type of incident, initial attack, and overhaul.

In the profession of fire fighting, there is no such thing as a routine call or an ordinary response. The evidence of this is in the number of firefighter injuries and fatalities that occur every year. "It seemed like a routine call . . .," begins many of the interviews with firefighters who have been involved in an emergency where a firefighter has been seriously injured or killed.

On emergencies, it is necessary to be aware of what is happening around you. While this seems obvious, it is not always as easy as it sounds. As you drag a hoseline into a house to extinguish a fire, work on the side of the highway with a critically injured patient, or cut a section of fireline on a wildland fire, it is not uncommon to become so focused in on what is immediately in front of you that you do not notice what is happening around you. This is commonly known as *tunnel vision* and most who work in the profession have experienced it at one time or another. Our vision can also become narrowed when an incident escalates, for whatever reasons, while we are in the process of trying to control it. This can lead to stress, fear, and panic. Ted Putnam[1] identifies and discusses how the thinking process can erode and regress under these types of conditions causing us to take fewer factors into consideration when making decisions. When this happens, having a crew that has trained together and worked together is an asset. Increased communication and receiving input from each crew member becomes increasingly important in recognizing all of the factors present. It is reported that it takes up to six weeks for a fire crew to form good crew cohesion. Until this takes place, there is a higher likelihood of accidents.[1]

Training in standard operating procedures with your crew and using SOPs on emergencies is important; however, this in no way implies that there are standard emergencies. While some incidents will have similar aspects, there is no such thing as a routine or standard emergency.

Questions

1. A primary search is made at a structure fire to determine
   A. the best strategy for extinguishment.
   B. if anyone is in the structure.
   C. if evidence of a crime exists.
   D. where the seat of the fire is located.

2. The establishment of the Incident Command System on hazardous materials incidents is required by whose authority?
   A. City
   B. County
   C. State
   D. Federal

3. What is the best approach to an emergency operation?
   A. Defensive
   B. Offensive
   C. Precautionary
   D. Aggressive

4. A vehicle fire should be approached from the
   A. front.
   B. front quarter.
   C. rear.
   D. rear quarter.

5. When fighting a fire involving a liquid propane gas container, how many gallons of water per minute should be applied to each area of flame contact by remotely supplied master streams?
   A. 150 gpm
   B. 250 gpm
   C. 300 gpm
   D. 500 gpm

6. A BLEVE produces a tremendously destructive explosion with container pieces traveling how far from the explosion site?
   A. 100 yards
   B. 500 yards
   C. half a mile
   D. 1 mile

7. When approaching a report of a structure fire, a self-contained breathing apparatus is necessary only when there is smoke showing.
   True   False

8. Power company personnel are the only fully trained and properly equipped people to deal with downed electrical lines.
   True   False

9. Firefighter turnout boots are made of rubber to repel water and insulate against electrical shock.
   True   False

10. On a wildland fire, the terms *backfire* and *firing out* are used interchangeably.
    True   False

11. When providing medical treatment, it is necessary to treat a patient who appears to be in good health with the same precautions as the patient who is obviously ill.
    True   False

12. The potential for a BLEVE is confined to LPG containers.
    True   False

13. When radiant heat from an oil tank fire presents a danger to other tanks, a master stream water curtain should be set up between the fire and exposures.
    True   False

14. It is not necessary to wear full structure fire PPE while extinguishing vehicle fires.
    True   False

15. Operations on a liquid propane gas leak without fire include controlling the leak and ignition sources and dispersing the vapor with a water fog.
    True   False

Match the term as it relates to oil tank fire fighting operations with the description that best describes it.

_____ 16. Boil over

_____ 17. Slop over

_____ 18. Froth over

A. Fire streams strike the liquid at an angle that pushes the burning liquid over the sides.

B. Water trapped below the level of oil heats up and expands to vapor at a rate of 1700 to 1 causing the oil to erupt outward.

C. Water turning to steam as it enters the burning liquid causing great disruption of the surface and the burning liquid to erupt outward.

19. The approach to any incident where there are suspected hazardous materials should be made from _____, _____, and _____.

20. The initial responsibility of the fire department at a hazardous materials incident is to _____, _____, and _____.

Exercises

Exercise 1

Read the information on the incident and answer the questions that follow.

On December 19, 1982, in Tacoa, Venezuela at 12:15 P.M., a storage tank, which supplied fuel for an electric generation plant, was the site of a violent explosion. The night before on December 18th, a high temperature indicator had alerted personnel that temperatures in a feed line from tank 8 were above normal. One of the two operating steam heating units in tank 8 was turned off and temperatures returned to within normal limits. The morning of December 19th, around 6:00 A.M., a three-person crew went to tank 8 for some routine maintenance. Two personnel went to the tank roof while the third waited in a vehicle inside the area of the tank dike. Within minutes, an explosion blew the tank roof off, set fire to the tank, broke lines that directed burning fuel into the dike, and damaged the fire protection system. Only the person in the dike was able to escape.

Fire personnel were called to the scene but were hampered by difficult access to the area that consisted of narrow winding roads, the number of spectators, and the damaged fire protection system. Once on scene, fire officials determined that the fire was contained to the tank. At that time, the ability to extinguish the fire was beyond the means available and the decision was made to let the fire burn itself out. Suppression support was then confined to a fixed monitor utilized by some of the firefighters.

At 12:15 P.M., with no apparent warning, an extremely violent explosion with fireball erupted from the tank. The fuel oil was thrown hundreds of feet into the air and flowed as far as 1300 feet from the tank. Many people were killed by the radiant heat and others by the hot oil. By the end of the incident, 150 people were dead, with 70 dwellings and over 60 vehicles burned. The number of people killed included 40 firefighters, 17 plant employees, media personnel, civil defense personnel, and spectators. Most of the fire apparatus at the incident was included in the number of vehicles burned.[2]

1. What is the term used to describe the extremely violent explosion which occurred during this incident?

   _____

2. In this type of incident, what events take place that result in an explosion?

   _____
   _____
   _____
   _____

3. Which operational mode of fire fighting was used on this incident?

   _____

4. When a fire is burning beyond the resources available to extinguish it, what steps can be taken to decrease the amount of life and property loss?

   _____
   _____
   _____

5. According to the textbook, what is the most effective agent and means of application for extinguishing oil tank fires?

   _____
   _____

Exercise 2

On July 6, 1994, at about 4:30 P.M., on Storm King Mountain in Colorado, 14 firefighters died when they were overrun by fire in the South Canyon Fire. Bruce Babbit, the Secretary of the Interior, had this to say about the incident, "This was an extraordinary tragedy involving some of the finest fire fighting professionals this country has produced."

The South Canyon Fire was started by lightning on Saturday, July 2nd. Because of the severity of the fire season, there were no local fire fighting resources available for initial attack on the fire. The fire was moving slowly through fuels of pinion juniper, oak brush, and short grass. A report from aircraft passing over the area on Monday was that there did not appear to be any active fire at that time.

On Tuesday, July 5, the first suppression resources were dispatched to the fire. These included three pumpers and 15 firefighters. That evening, a crew of smokejumpers was dropped on a ridge near the fire. Because of mechanical problems with chain saws, the IC, by radio, turned the fire over to the smokejumper in charge and left the fire with the crew. The fire continued to burn and grow throughout the night, and additional resources were requested by the smokejumper in charge of the fire. On Wednesday, July 6, additional resources were committed to the fire. They included 20 firefighters, two aircraft, and ground personnel. The three crews on the fire that consisted of 47 personnel, hot shots, and smokejumpers, were divided into two groups, with another group of nine smokejumpers breaking off from those groups.

Each of the three groups then worked to secure their part of the fire line in the steep, rugged terrain. Firefighter Brad Haugh noted, "The fire was just creeping downhill. It was an average fire with bad access and rough terrain."

Around 2:30 P.M., fire activity began to increase in the pinion juniper. About 3:30, a cold front began moving into the area causing gusty and unpredictable winds. Helicopter water drops were called in to control the flareups on the fireline. About 15 minutes later fire was spotted starting up the bottom of the drainage and ridge. This probably was caused by burning material rolling out of the fireline and down the hill. Crews were given the word to head for designated safety zones. The blowup occurred rapidly. Derek Bixley describes it this way, "We saw the smoke build up, and then WHAM!—the whole mountain was on fire."

With the deaths of 14 highly trained professional firefighters came in-depth investigations of the South Canyon Fire. These investigations have resulted in the awareness that there are more lessons to be learned and more training to be accomplished. Some of these areas include: increased training in the use and deployment of fire shelters during intense fire behavior; reinforcement of fire orders and "Watch Out" situations; and training in decision making during periods of extreme physical and psychological stress.[1,3,4]

1. Compare the four most common causative factors involved in tragedy and near-miss wildland fires and identify if and how any of those factors existed at the incident.

   _____
   _____
   _____
   _____
   _____
   _____

2. Using the 18 Situations That Shout Watch Out, identify those present in the incident.

   _____
   _____
   _____
   _____
   _____
   _____

3. What does LCES stand for?

   L: _____

   C: _____

   E: _____

   S: _____

4. Review the Ten Standard Fire Fighting Orders and briefly describe how they relate to an incident of this type.

_____

_____

_____

Exercise 3

On hazardous materials incidents, perimeters are set up to control scene access. Label the three zones of the hazardous materials incident found in Figure 14.1.

1. _____

2. _____

3. _____

What type of personal protective equipment would emergency workers be wearing in each zone?

1. _____

2. _____

3. _____

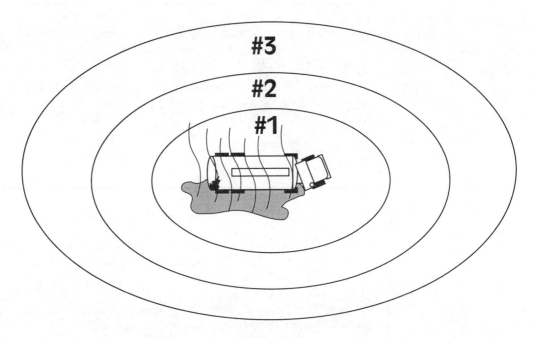

Figure 14.1

Assignments

Answer the following questions as they relate to the local fire department or the fire department in the jurisdiction of your choice.

1. What is the initial structure response by the department?

    Number of Engines: _____

    Number of Personnel, list by rank: _____

    _____

    Support Equipment: _____

    Support Personnel: _____

2. What is the initial wildland fire response?

    Number of Engines: _____

    Number of Personnel, list by rank: _____

    _____

    Support Equipment: _____

    Support Personnel: _____

3. What types of mutual aid agreements does the department have with surrounding areas?

    _____

    _____

4. Does the department have a full-time hazardous materials response team and support vehicle? If so, indicate the number of personnel and equipment. If not, where does the hazmat unit come from when needed?

    _____

    _____

NOTES

1. Putnam, Ted (September, 1995). The collapse of decisionmaking and organizational structure on Storm King Mountain, *Wildfire*.
2. Henry, Martin & Klem, Thomas (June, 1983). Scores die in tank fire boilover, *Fire Service Today*.
3. Putnam, Ted (September, 1995). Analysis of escape efforts and personal protective equipment on the South Canyon Fire, *Wildfire*.
4. Newcomb, Charles (August/September, 1994). *Firestorm claims 14 in Colorado*, Nassau, DE: Firefighters News.